The Finger Lakes Region

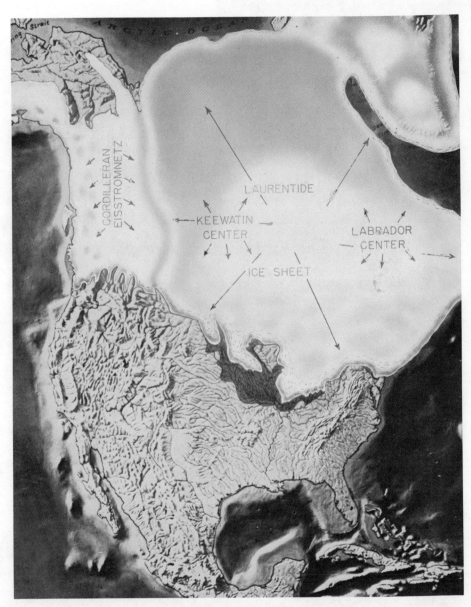

Continental glacierization of North America in the Pleistocene epoch. The light areas show the maximal extent of the ice in the latest advance; the dark area to the south indicates the farthest extent of an earlier advance in the same epoch.

The Finger Lakes Region
ITS ORIGIN AND NATURE

By O. D. von Engeln

CORNELL UNIVERSITY PRESS

ITHACA AND LONDON

Copyright © 1961 by Cornell University

All rights reserved. Except for brief quotations in a review, this book, or parts thereof, must not be reproduced in any form without permission in writing from the publisher. For information, address Cornell University Press, 124 Roberts Place, Ithaca, New York, 14850.

First published 1961 by Cornell University Press.
Third printing 1969.
Reissued 1988.
First published Cornell Paperbacks 1988.

Library of Congress Cataloging-in-Publication Data

Engeln, O. D. von (Oscar Diedrich), 1880-1965.
  The Finger Lakes region : its origin and nature / by O.D. von Engeln.
           p.   cm.
Bibliography: p.
ISBN 0-8014-0437-1 (alk. paper)
ISBN 0-8014-9501-6 (pbk. : alk. paper)
  1. Geology—New York (State)—Finger Lakes. 2. Physical geology—New York (State)—Finger Lakes. 3. Glacial epoch—New York (State)—Finger Lakes. I. Title
QE146.F6E5 1988   551.4′09747′8—dc19   87-34011

Printed in the United States of America

The paper in this book is acid-free and meets the guidelines for permanence and durability of the Committee on Production Guidelines for Book Longevity of the Council on Library Resources.

FOR

The Faculty Stalwarts: Tarr, Gill, Harris
of the Good Old Days in the
Geology Department of Cornell University

ACKNOWLEDGMENT

The author is highly appreciative of the financial aid received through a Faculty Research Grant, Cornell University, awarded him to defray expenses incurred for drawings, photographs, and clerical services in the preparation of the manuscript of this book.

★ Contents ★

	Prologue	1
1	Escarpments and Cuestas	11
2	The Peneplain Uplands and Their Dissection	32
3	Finger Lakes "East"— Finger Lakes "West"	38
4	Glacial Erosion and Rock-Basin Lakes	46
5	Hanging Valleys and Through Valleys	56
6	Rock Gorges and Two Glacial Advances	69
7	The High-Level Finger Lakes	86
8	Glacial Deposits	102
	Epilogue	121
	Appendix I: Vantage Points and Excursions	131
	Appendix II: Maps	137
	Glossary	139
	References	149
	Index	153

⋆ *Illustrations* ⋆

Continental glacierization of North America	*frontispiece*
Columbia Glacier, Alaska, advancing into forest	*facing page* 4
Vegetation at the side of Shoup Glacier, Alaska	5
Lower end of gorge, Watkins Glen	20
Peneplain uplands	21
Keuka Lake	36
Glacial erosion by grinding	37
Berby Hollow	52
Sixmile Creek through valley	52
Labrador Hollow through valley	53
Postglacial consequent gorge	68
Interglacial consequent gorge	*between pages* 68–69
Rock pillar, Buttermilk Gorge	68–69
Fall Creek at Forest Home in flood	68–69
Ithaca Falls, Fall Creek	68–69
Taughannock delta	68–69
Taughannock Gorge rock fall, 1935	68–69
Taughannock Falls in 1888	68–69
Taughannock Falls	68–69

Cavern Cascade, Watkins Glen 68–69
Cascadilla Creek gorge 68–69
Chute channel in Enfield Glen . . . . . . *facing page* 69
Coy Glen terraces 84
Cayuta Lake outlet gorge 85
Ice cliff of a glacier 85
Proglacial lake succession 100
Drumlin 101
Moraine soil with angular boulders 101
Glacial boulder 116
Wasting glacier 116
Moraine topography, Cayuga Inlet valley 117
Lake clay 132
Kame deposit near Freeville 132
Esker at Malloryville 133
Index map for the United States Geological Survey maps . .
between pages 148–149

DRAWINGS

Cuesta development . . . . *page* 12
Rock belts of central and northern New York 14
Hooked tributary, Salmon Creek 21
Preglacial Cayuga "River" drainage 24
Postglacial Cayuga drainage 25
Finger Lakes "West" 28
Finger Lakes "East" 29
Connecticut Hill 34
Cayuta Lake outlet gorges, 1893, 1944–1950 82
Glacial Cayuta Lake, first stage 83
Glacial Cayuta Lake, second stage 84
Fossil waterfall at Clark Reservation 98

The Finger Lakes Region

Prologue

IF, like Peter Bell,[1] you are quite insensitive to the beauties of Nature, and incurious about how natural phenomena originate, the content of this book will hardly merit your reading effort. If, on the other hand, you enjoy picturesque and imposing scenery and have, besides, a lively curiosity about the manner and the sequence of events pertaining to the development of landscape features, the perusal of this account of the Finger Lakes Region of Central New York should prove definitely rewarding.

Such recompense will derive primarily from the circumstance that the region is scenically unique. Although the residents of the area, no less than tourist visitors, are well aware of the attention-arresting nature of many aspects of the Finger Lakes landscapes, probably few persons in either of these groups realize how completely distinctive they are.

In a geographical environment that has as its general expression wide farm acres and level upland pastures, eleven long, narrow,

---

[1] A primrose by a river's brim
A yellow primrose was to him
And it was nothing more.

roughly parallel lakes appear. These lakes extend south to north as might the fingers of a pair of outstretched hands. The southern ends of the lakes have high, almost cliff-steep, shores. Two of the lakes are so deep that their bottoms are below sea level. Associated with the lake basins are narrow, deep gorges, with cascading and plunging waterfalls. South of the lakes great troughs with nearly vertical rock walls extend through the uplands for miles without interruption. These are scenic phenomena appropriate to rugged mountains and high plateaus but are singular features where encountered in a placid countryside.

While such an association of land forms is a quite anomalous development, hence, of itself, provides a warrant for designating the Finger Lakes Region a topographically unique occurrence, the manner of its origin is, further, without parallel elsewhere in all the wide world. Here a major unit of surface relief was opposed to the onset of an earth force of gigantic magnitude. The magnificent demonstration of the immovable object and irresistible force issue that ensued resulted in the fashioning of the completely distinctive land forms and scenic features of the Finger Lakes Region.

This climax was preceded by a succession of great geological events which, in sequence, required an exceedingly long period of time to run their course. The rest of this Prologue is devoted to a brief review of this sequence and should serve to set the stage for the topics of the chapters that follow.

The geologic history of the Finger Lakes Region began about 550 million years ago, when much of the eastern area of North America was a nearly featureless plain underlain by crystalline rocks.[2] There were highland tracts, but with the exception of part of the Adirondacks all of New York State appears to have been a low-lying, gently undulating expanse of land with altitude only slightly above sea level. How long this low-lying surface had existed is

---

[2] These determinations of geologic time are made by computing how much of a radioactive element, such as uranium, in the rocks has been converted to lead by radio emissions since the rock was formed. With modern techniques such calculations of geologic time are fairly accurate in round numbers.

Prologue

indeterminate, but at the 550-million-years-ago juncture the region sank so much that the sea spread across it. This submergence continued for 325 million years. Throughout this very long interval waste products, debris resulting from the crumbling of the ancient rocks of the Adirondacks and of high lands in New England, were washed down by rivers into the shallow sea covering the territories to the south and west of them. These materials were fragments resulting partly from the mechanical breaking down of the rocks, primarily sand and pebbles, and partly from their chemical decay, mud and lime. The lime is carried away in solution.

Although conditions did not remain completely uniform during all this time (for example, there was a period many millions of years long when the ocean waters collected in vast lagoons and evaporated in such volume that layers of salt hundreds of feet thick were laid down), the net result was the accumulation of bed after bed of sand, mud, lime, and salt, one above the other, all in a gently inclined position sloping away from the ancient shore lines. The bottom of the sea must have sunk progressively to make room for the accumulating beds of sediment. Eventually, from the bottom upward, the layers were hardened into rock and became sandstones (sand grains cemented together), shales (compacted mud or clay), and limestones (composed of lime carbonate). In the Finger Lakes Region the pile was found to be 8,000 feet thick by a boring which penetrated to the crystalline rocks (made up of an aggregate of mineral crystals) of the original sea bottom.

The occurrence in practically unbroken succession, and now largely exposed to view, of beds representing this enormous lapse of ancient geologic time is another aspect in which the Finger Lakes Region is unique. Nowhere else in the world is such a section present.

About 200 million years ago a vast uplift of the earlier sea bottom occurred. In the Finger Lakes Region and in the lands to the north and west of it the attitude of the beds remained essentially unchanged from that of their deposition on the sea floor. Farther south, however, in Pennsylvania, the Appalachian Mountains were first raised at this time. There the great pile of beds was laterally compressed into gigantic corrugations, in form like a series of huge smooth waves approaching the shore from the open sea.

Not improbably the sea bottom first became dry land around the bases of the Adirondack and New England high lands which had furnished the sediments deposited on it. As the beds emerged, their inclination was steepened, but they preserved a general downward slant (regional dip) toward the south. Drainage on the new land surface flowed down this original slope and eventually was organized into trunk streams, now represented by the Susquehanna and Delaware river systems. It would seem that the Appalachian corrugations, by rising athwart the course of this drainage toward the sea, would have blocked its flow. Instead the rivers simply sawed (eroded) through the upbends of the rocks as fast as these rose across their paths.

There ensued a long period during which the rock mass, tilted beds and corrugated beds alike, was under the sky and subject to the attack of rain, frost, and rivers. This destruction of the uplifted rocks proceeded until all the New York country and territories far outside it were worn down to a surface of very low relief nearly at sea level. A land form so produced is called a peneplain, a coined word meaning almost a plain. This would seem to be almost an incredible measure and completeness of downwearing, for the altitude of the uplifted lands must have been 3,000 feet or more. But evidences of such downwearing have been preserved. Thus, in Central New York instead of the tops of the inclined beds being the surface of the land, it is their beveled edges which appear as a series of parallel bands extending east-west across the state. In the Appalachians the upfolds of the beds have their tops sliced off as if by a huge knife. The evidence for this slicing off is that the summits of the present-day Appalachian ridges, despite differences in the nature of their rocks, are planed down to matching levels.

Although a peneplain is worn down by erosion to near sea-level altitudes, it is not a flat surface; sufficient gradients exist for streams to flow. In the New York area south of the Adirondacks the drain-

Columbia Glacier, Alaska, advancing into forest. This illustrates how the front of the continental glacier advancing across Central New York overwhelmed the forest and plowed up the soil at its front. (Photograph by O. D. von Engeln.)

Vegetation, Temperate Zone kinds, at the side of the shrinking Shoup Glacier, Alaska, indicating the mild climatic conditions that existed at the margin of the continental glacier when it was melting off New York State. The ice mass of the glacier is on the right. Rock fragments melted out of the ice are building up ablation moraine. These rock falls prevent the vegetation from growing right up to the glacier ice. (Photograph by O. D. von Engeln.)

age persisted in its southerly direction to the Atlantic. Farther west the same southerly direction of flow obtained from a line in general south of the outlets of the present Finger Lakes. But north of this line the peneplain runoff may be presumed to have been toward Hudson Bay. In other words, a low water parting along a line running southwestward from the southwest base of the Adirondacks was developed on the peneplain. The existence of this divide on the peneplain was a circumstance of basic significance in the later evolution of the land forms of the Finger Lakes Region.

The period of stable land level which permitted the production of the peneplain endured for about 100 million years. It was brought to a close by renewed upheaval. This second uplift of the land was in the nature of an upbulging. The drainage established on the peneplain persisted although the axis of highest uplift was again athwart the courses of the south-flowing streams. South of the axis of uplift the gradients of the streams were steepened, and this increased their erosive powers sufficiently to permit them to saw through the bulge as fast as it rose. Thus the Delaware and Susquehanna water gaps came into being.

The courses of the postpeneplain trunk drainage northward from the peneplain divide are not now discernible. One development, however, is clear. The divide was raised to a high altitude, and no barrier was opposed to a descent from the new heights to the Arctic seas. Further, the north-flowing streams had to cut through only the thickness of such beds they encountered that were resistant to river erosion, whereas a south-flowing stream had to saw lengthways through every durable bed its erosion uncovered.

This difference in conditions affecting the north- and south-flowing streams governed the later land-form development of the Finger Lakes Region.

The major north-flowing streams rapidly cut notches across the outcropping edges of resistant beds. This notching permitted effective deepening of the channels of their lower courses across easily eroded beds. Such deepening, in turn, greatly steepened their upper courses. Steep descents, there, facilitated the gnawing back of the headwaters across the peneplain divide after the manner that a rain gully lengthens its course across a newly graded slope when subject to a thundershower downpour. Thus the north-flowing streams gained in volume by reversing the flow of the upper waters of the south-flowing streams.

A bed or formation designated as resistant must, to be given such distinction, be overlain and underlain by weaker strata. In the Central New York area the weak formations are exceptionally thick in the Finger Lakes north-south belt and in the country to the north of this belt. East-west tributaries to the trunk north-flowing streams located on the weak formations were given especial erosive competence as they rushed down to the deepening cuts. Short tributaries to the east-west streams were, in turn, quickened so that in a relatively short time the whole width of a weak belt was reduced to a lowland. Theoretically this lowland could be cut down until its floor had descended to the top surface of the next underlying resistant formation. By the headwater erosion process, also, the heads of the east-west tributaries to the trunk north-flowing streams encroached on drainage areas formerly tributary to lesser north-flowing streams. In time wide reaches of a lowland belt were all linked up in a single east-west system tributary to a large, or the largest, north-flowing trunk stream.

The postpeneplain stream erosion history outlined above continued without interruption for so long that the uplifted peneplain surface was deeply dissected. North of the peneplain divide the sculpturing erased all traces of the earlier nearly level land form. Broad valleys with gentle gradients linked up in completely organized and stable drainage systems comprised the relief. Beyond the erstwhile divide the streams of the south-flowing drainage, despite the structural handicap imposed on their erosional activities, had carved deep well-defined valleys leading down to the great water gaps across the belt of folded rocks.

But about one million years ago this placid and orderly sequence of landscape evolution was abruptly terminated when the greater part of northern North America was occupied by glaciers of con-

Prologue

tinental magnitude. In other words there came the time of the Ice Age.

A first view of a glacier, if only of a small one, is an impressive experience. It is, indeed, even *awesome* to see so much *ice* concentrated in one mass and existing, as it were, in its own right. If the observer's first contact is with a really large glacier, his reaction is akin to that evoked by realization of the limitless expanse of the sea viewed in mid-ocean from the high deck of a ship. With this comparison in mind it is possible, faintly, to conceive of the overwhelming mass of the continental glaciers that spread over North America during the Ice Age.

The Ice Age continental glacier of eastern North America originated as a low mound of ice on the upland of the Labrador Plateau. This mound grew in thickness until it attained a depth of 10,000 feet in the Labrador source area. Meanwhile it spread laterally until in the line of the Finger Lakes the ice extended southward to Williamsport, Pennsylvania. In the Central New York lowland the ice was 2,500 or more feet thick at its maximum stage.

The advance of the glacier front from its originating area was slow, on the order of two or three feet per day, but inexorable. All plant and animal life was obliterated or compelled to flee southward when the ice came. The expansion of the glacier was, however, not a mere sliding forward of the ice. A crustal cover, about 200 feet thick, of rigid ice was carried along on the main mass of the glacier (ten or more times as thick below the rigid cover), which moved by viscous flow. Viscous flow, as here conceived, substitutes the individual crystals making up the glacier for the molecular or atomic particles involved in viscous flow as conceived by the physicists. The ice crystals are surrounded by a liquid film which facilitates intercrystal movements of rotation and translation. In sum these displacements are converted to the movement of the whole glacier mass at the rate of 3, 10, or even 25 feet per day. The motion is steady and continuous. During advances of the glacier the forward flow of the ice at depth is speeded up by progressively augmented supplies from the Labrador source region. Thus not only is the rate of glacier motion maintained but a sufficiently greater volume of ice is furnished to expand the glacier-covered area.

There is currently a misconception, on the part of the younger generation of glacialists, that the ice front was bordered by a zone of bitterly cold climate. Actually the temperatures of the area bordering the front were probably little different from those of today at the same latitudes. Polar cold may have existed in the interior of the ice sheet but its terminal margin represented at every stage a balance between supply and melting. The floods of water that were released at the ice margins when the glacial epoch came finally to a close are clear evidence that the ice was not dissipated through dry, cold evaporation. The conditions instead were comparable to those existing today at the borders of the large ice fields of Alaska and Iceland.

The general direction of the advance of the glacier was southward. At the north end of the Central New York lowland the ice, in effect, entered the broad mouth of a land bay which narrowed and was rimmed by uplands at its southern end.

As the ponderous ice mass advanced across country, it exerted tremendous erosive power on the rocks of its bed. The bottom of the glacier, shod with rock debris progressively acquired in the southward flow, was a gigantic abrasive tool pressed down by the weight of hundreds of tons of overlying ice. The glacier also exerted a quarrying action on the bedrock, wedging out blocks of all sizes, many of enormous dimensions. Some of these were shunted up into the body of the glacier by rising internal currents of ice flow. Other large rock fragments were detached and incorporated in the upper levels of the ice when the glacier flowed around isolated summits in the Laurentian and Adirondack regions. No matter what their size all such big blocks were carried forward as effectively and steadily as the finer stuff and mixed in with the finer materials when finally deposited.

A notable phenomenon of these eroding and transporting processes of the glacier is that the interior and bottom ice flowed in currents which were significantly channeled by the relief of the country passed over. Thus the Central New York lowland with its wide opening in the north acted as a funnel in directing and progressively constricting the southward flow of the ice. The large volume of ice entering the wide funnel mouth could originally be accommodated in the narrowing southern section only by accelerated, localized rate of flow. Faster currents were thus engendered along the valley lines oriented in the direction of the advance. Accordingly, the erosive effectiveness of the glacier was greatly enhanced in these channels. The eastern Finger Lakes rock basins are the conspicuous production of glacially overdeepened channelways.

Although hundreds of millions of years of geological time (preceding the Ice Age) are recorded in the rocks of the Central New York lowland the accumulated beds contain no evidence of an epoch of continental glacierization in all that interval. It is, accordingly, quite contrary to expectation to find that within the million years or less of the Pleistocene epoch (the Ice Age) there were two continental glacierizations of eastern North America, especially since these ice advances were separated by an interglacial time during which the climate was warmer than that of the present. Indeed, in the Middle West and in Europe, four advances and intervening deglaciations are well authenticated, all within this relatively short geological time of one million years. In Central New York, however, there is clear evidence of only two glacial occupations with an interglacial interlude, longer than postglacial time, coming between.

In the Finger Lakes Region, moreover, the major erosional modifications of the rock topography were made by the *first* glacial invasion. The second ice advance, in effect, found the way prepared for it; it fitted into the grooves fashioned by the first glacier.

Contrariwise, the other great change in the aspect of the countryside affected by glaciation was due to the *second* ice advance. This change resulted from the deposit of enormous amounts of rock debris when the second glacier melted off the land. By contrast with the dynamic nature of the advance of the ice its disappearance was static. When not enough ice was brought forward to maintain the front of the glacier along a given terminal line, the unnourished areas of the glacier simply wasted away and the active terminus was established along lines progressively farther north. The deposits of rock waste laid down as the ice melted were of two types: (1) materials deposited directly from the ice, debris that had been carried on the surface of the glacier, in the body of the ice, and at the sole of the glacier, such as boulders and rock flour, and (2) materials transported to the glacier front and carried away from the terminal line by the floods of water released by the ice melting, such as gravel, sand, and clay. Both types of material during uniform recession of the glacier front were spread widely and evenly, but when equilibrium between supply and wasting was maintained for a considerable period (100 to 200 years) they were concentrated in the localized terminal accumulations that are topographically conspicuous today.

Postglacial time in the Finger Lakes Region extends over some 9,000 to 10,000 years. During these years erosion by running water has wrought notable sculpture changes in the landscape. The innumerable rock gorges (of which Watkins Glen is the best known), which together with their waterfalls are so characteristic a feature of the picturesque Finger Lakes Region scenery, were cut in this time and, in fact, are still being enlarged and deepened.

Such, in brief, is the geomorphic history of the Finger Lakes Region. The prologue provides the background essential for understanding and appreciating the distinctive landscape forms of the area.

Interpretations of specific phenomena are the topics of the chapters which follow. These interpretations are presented as the product of fifty years of professional study and field research. In considerable part they have not been heretofore publicized except as class instruction. The author has endeavored to make the discussions informing for both the scientific reader and the intelligent layman.

1 · *Escarpments and Cuestas*

WHERE thick accumulations of layered sedimentary rocks—shales, sandstones, and limestones—exist, as in Central New York, and these beds have been lifted high above the sea floor on which they were originally deposited, they may be disposed in one of three possible attitudes: (1) they may retain the nearly horizontal position they had on the sea bottom; (2) they may have been caused to depart from such original horizontality to the extreme of standing vertically on their edges; (3) they may merely have been given a moderate inclination.

When, after uplift, such a layered pile of rocks is subject to sculpture by rain and rivers through long ages of geologic time, imposing and distinctive landscape forms result. If the horizontal bedding has been retained, spectacular gorges, cliffs, buttes, and mesa-plateaus are created, especially in arid regions, by the erosional attack. Such a result is exemplified by the scenery of the Grand Canyon of the Colorado and the regions adjacent to it. Where in complete contrast the beds stand on end, narrow, sharp-crested, vertical or near-vertical ridges, popularly called hogbacks, emerge. These are composed of the strata that are particularly resistant to the destructive processes of nature. Hogbacks on a grand scale flank the eastern slopes of the Rocky Mountains in Colorado and ring around the Black Hills in South Dakota. Where, as in Central New York, the beds neither are flat-lying nor stand up vertically but have instead a moderate inclination (regional dip), the effect of long subjection to weathering and stream erosion is no less characteristic than the mesas and hogbacks, though scenically

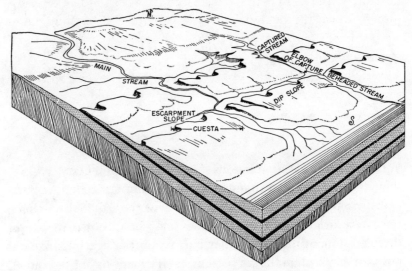

Cuesta development. The diagram shows how a major stream (e.g., the preglacial Cayuga "River") extends its length by headwater erosion in gently dipping strata and how lateral tributaries open wide valleys in weak rock bounded by escarpment cliffs held up by strong beds (black). The escarpment cliff and the gentle back slope jointly constitute a cuesta. (Diagram by Audrey H. O'Connor.)

not so spectacular. The inclined structure gives rise to alternations between cliffs or steep slopes and wide gentle slopes. The steep slopes mark the lines where the edges of the strata terminate at the surface, the gentle slopes are their dipping top surfaces. The eventual over-all landscape development is the creation of parallel rows of hills each with one steep slope and one gentle slope. These hills are designated by a Spanish term, *cuesta,* which signifies a hill of such profile.

Escarpments and Cuestas

The sediments that comprise the strata of the Central New York region were derived from the Adirondacks and other highlands to the northeast and east of the area. Each bed, while it was being deposited, must have terminated at the shore line where this lapped around the base of the source country. Accordingly, the original ends of the successive layers were on the north and east sides of the area that was covered by the deposits. Since the trace of the shore line, because of changes in the level of the sea and of the height of the land, shifted from time to time during the geological ages through which the deposition was in progress, the accumulation and spread of the beds were not quite so simple and coextensive as this outline would suggest.

Eventually the ages-long sinking of the bottom of the inland sea, which had permitted the building up of this enormous thickness of sediments, was reversed. Instead of sinking, the whole area began to rise and was never thereafter submerged. A feature of this uplifting was the creation of the Appalachian Mountains of Pennsylvania. In that area the pile of beds was uplifted by being forced, through lateral thrust from the southeast, into great folds much in the manner that a blanket wrinkles when it is pushed toward the foot of the bed.

North and west of this belt of corrugated beds there was a bodily upheaval of the strata. The uplift, in other words, was vertical, so that the new land surface was a plain slanting southward toward the folded belt and toward the sea. The greatest heights were attained at the base of the source regions—e.g., in New York, the Adirondacks. From these, accordingly, major streams flowed over the plain and across the mountain belt down to the Atlantic. The downcutting of the streams more than kept pace with the rise of the upfolds.

There followed a long period, millions of years, of stability of level. During this time all the region of the sedimentary-rock accumulation, fold mountain and plateau uplift alike, was worn down by decay, frost breaking, and river wear practically to sea level, producing a gently sloping plain, or peneplain. Since the uplift was perhaps no greater than 2,000 feet, there remained still a mass of sedimentary beds 6,000 or more feet thick below sea level in the central and southern parts of the original area of deposition.

Around the northern border of the area, however, where the contact with the source region of the sediments existed and where the measure of the uplift was greatest, the peneplanation brought about a particular and significant pattern of rock outcrop. Since the beds had retained their gentle southerly inclination, wearing them down vertically beveled across their overlapping terminal parts so that these were exposed as a series of wide belts extending east west across country. The thicker a given formation, the wider the belt it gave rise to on the surface. If the land surface is horizontal, a bed 100 feet thick dipping 3 feet to the mile would give rise to an outcrop (surface exposure of the bed) 33 miles wide from north to south. Accordingly, if after peneplanation conditions ne-

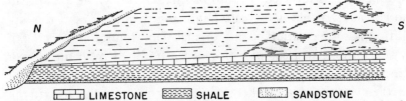

LIMESTONE   SHALE   SANDSTONE

Rock belts of central and northern New York, illustrating how the level peneplain surface, underlain by very gently inclined strata gives rise to broad (north to south) belts of country, each being the surface expression of a single rock formation. (Diagram by Audrey H. O'Connor.)

cessitating renewed active erosion were introduced, each of the different formations, alternately durable and weak beds, was immediately available to be acted upon by the attacking agents and would yield in accordance with the degree of its resistance to the destructive processes.

These conditions were provided about 30 million years ago when the peneplaned area experienced a bodily uplift of some 2,000 feet. This second rise affected the drainage developed on the peneplain. This peneplain drainage pattern differed from that developed on the original uplift in that a low divide had become established on the peneplain extending southwestward from the southwest corner of the Adirondacks. Streams north of this water parting flowed toward the Arctic Ocean. But to the south and east of the new divide the southeast drainage of the initial uplift persisted.

While the second uplift did not affect the pattern of the southeast drainage, it did bring about a renewal of active downcutting by both the trunk streams and their tributaries. The great valleys of the Susquehanna and the Delaware are the products of this postuplift erosion. But the increased gradient given to streams flowing northward from the divide on the peneplain surface resulted in the creation of a quite distinctive topographic development. A cuesta landscape of diagrammatic perfection emerged in consequence of the erosive dissection of the sequence of formation belts exposed there.

Although no trace of them remains, it may be assumed that the trunk streams north of the peneplain divide flowed toward the Arctic Ocean. They would, accordingly, cross in turn, practically at right angles, each of the belts of the outcropping formations encountered north of the divide. The belts of weak rock were readily trenched to a depth governed by their thickness. Since the weak formations were regularly much thicker than the resistant ones and their belts of outcrop correspondingly wider, a considerable drainage could collect on their surfaces. Such drainage would be tributary from both the east and west to the north-flowing trunk streams. Since the weak belts, owing to their thickness and low-angle beveling, were miles wide, the tributary streams developed on their surfaces had ample scope for, and rapid yield to, their erosional activities. The deep trenches made by the major streams steepened the gradient of the tributaries. Taken together these favorable conditions brought about the speedy excavation of the whole bulk and thickness of the weak formations.

In this process the downward cutting of a tributary went forward without a check until the upper surface of the next underlying resistant formation was encountered. Then a new factor became operative. Because of the southward dip of the strata, the top of the resistant bed would be lower on the south side of the tributary. Accordingly, the tributary could cut deeper there in the weak material, and the most water and the strongest current would concentrate there. As a consequence, the stream would very effectively erode sideways as well as downward. Such lateral shifting of a stream course is called migration down the dip.

In time this phenomenon had noteworthy topographic consequences. The continuous gnawing away at the south bank by a tributary stream resulted in the creation of a cliff or sharp slope whose height was determined by the vertical thickness of the weak bed plus the thickness of the resistant bed overlying it. Thus the escarpment slope of a cuesta is caused to emerge. It faces the wide gently inclined dip slope across which the stream course has migrated.

Although no recognizable evidence exists of the course of the postpeneplain master streams flowing north to the Arctic, not improbably the middle and lower sections of such original trunk streams were in time diverted to Atlantic Ocean or Mississippi River outlets through progressive linking up of east-west streams developed on the weak rock belts at the base of cuesta escarpments. The position and extension of the basins of Lake Erie and Lake Ontario are possibly relics of a pattern of drainage so organized. Farther south the axes of Cayuga Lake and its headward valley prolongation are apparently in the line of the longest and largest of the preglacial north-flowing streams in the Finger Lakes Region.

The effacement and reconstruction of the north-slope postpeneplain drainage system was brought about by the Ice Age glaciers. By erosion and deposition the ice advances and retreats rather completely disrupted preglacial stream courses north of the divide. On the other hand the glacial visitations failed to erase completely the cuesta topography that was the dominating element of the preglacial relief in the sedimentary belts. Although obscured and breached in places, the escarpments remain the basic element of relief in the present land-form expression and together with the dip-slope lowlands, concurrently developed, are the major units of the regional landscape. These are not necessarily the most conspicuous and eye-catching scenic forms, but they reflect the fundamental structural backgrounds of the present-day topography.

Escarpments and Cuestas

The critical units in cuesta development are the resistant formations. In the Finger Lakes Region and the bordering lands on the north and south there are four of these.

Farthest north is the Niagara limestone (technically the Lockport dolomite), the formation which gave rise to the Niagara escarpment. This was probably the first escarpment developed in the postpeneplain period of dissection and still has conspicuous and rather continuous expression. Where it is prominently developed, as at Lewiston, New York, it is the outstanding feature of the local relief. The Niagara limestone is a massive bed, 140 to 150 feet thick. It is underlain by the weak Rochester shale, 100 feet thick. Below this are other soft shales. This succession of very strong and exceptionally weak formations provided ideal conditions for escarpment and lowland development. The lowland is the Ontario Plain created by the excavation of the underlying shales and the associated ages-long retreat southward of the escarpment front. There should be a great east-west river flowing along the base of the escarpment, but if this was once present little trace of it now exists. Its presumed course is largely occupied and blanketed by deposits of debris brought by the ice of the Pleistocene glacial epoch. There is a buried valley connecting Georgian Bay and Lake Ontario whose rock bottom is far below lake level. The line marked by the base of the escarpment is also followed by the Trent River, which, supplemented by canals, permits navigation from Lake Ontario to Lake Huron.

The second (next south) escarpment is fixed by the Helderberg and Onondaga limestones. The Helderberg escarpment overlooks the Mohawk Valley and is a cliff rising 900 feet above the valley floor. (Farther east, where the escarpment faces the Hudson Valley, the cliff constitutes the Helderberg Mountains.) Its accompanying lowland in the west is the wide Mohawk Valley.

West of Utica the Helderberg limestone thins and is replaced by the Onondaga limestone, which is higher in the pile of beds than the Helderberg formation. In western New York the Onondaga is 100 to 150 feet thick. The descent from the south over the Onondaga at Syracuse is an abrupt declivity of about 100 feet. Farther west the Onondaga escarpment is topographically inconspicuous. At Waterloo and Seneca Falls, however, where the Seneca and Cayuga Canal crosses the Onondaga outcrop, there is a descent of 49 feet necessitating passage through three locks to reach the lower level. The general line of Onondaga outcrop is at the northern ends of the larger Finger Lakes.

Despite the present slight relief of the western extension of the Onondaga escarpment, the lowland at its base has a significant expression. It determines the lower course of the Seneca River and, farther west, that of lesser west-east streams. It is the site of a wide belt of swampy tracts. The patternless drainage courses of these parts of the Onondaga lowland may be attributed to the uneven thicknesses of the glacial deposits which blanket a rock surface presumably of very low relief, namely that of the dip slope toward the base of the escarpment.

The third distinctly resistant formation, the Tully limestone, crops out south of the Onondaga escarpment, on a line extending from Cazenovia on the east to Seneca Lake. Although massive, this limestone is only 15 to 20 feet thick, and if, preglacially, it had given rise to a conspicuous cliff this must have been reduced by glacial erosion and obscured by glacial deposits along much of the outcrop line. A Tully escarpment does, however, attain some measure of prominence south of the broad lowland extending from East Homer to Truxton, and there the East Branch of the Tioughnioga River flows parallel to its base. Since the line of the Tully edge is south of the north ends of the eastern Finger Lakes, escarpment development due to it is interrupted by the gaps made by the lake valleys. Between Skaneateles and Owasco lakes the Tully lowland is marked by the broad Mandana trough, now practically devoid of drainage, but that apparently was an important east-west stream valley before being cut off at both ends by the creation of the lake basins.

The southernmost escarpment is called the Portage. Portage is a formation name formerly used by stratigraphers to designate the succession of shale and sandstone beds of what is now called the Finger Lakes stage of sediment deposit. In the Finger Lakes stage and the Chemung stage, which is next above it in the rock column, durable sandstone beds occur and become increasingly numerous in the upper levels of the pile. Collectively these sandstone strata, especially the thicker ones of the Chemung stage, serve to maintain the Portage escarpment front. At best, however, their functioning as capping layers is cumulative rather than single and decisive along one fixed line, as is the case with the other escarpments whose occurrence is determined by massive limestone unit beds.

Despite this lack of a specific crest-determining layer, the Portage escarpment is now the dominant over-all relief element in the topography of the Finger Lakes Region. Where it is characteristically developed, the Portage front is 900 to 1,000 feet high. Because of its structural make-up it is susceptible to piecemeal destruction and regularly appears as a steep slope, never as a vertical cliff. Although composed of discontinuous, sometimes repetitive, parallel units, the Portage escarpment is nevertheless persistent in occurrence, and this persistence indicates the major reason for its existence, namely that it marks the northern edge of the relics of the uplifted peneplain surface at, roughly, the line of the postuplift divide between the north-flowing and south-flowing drainage. That this divide line happened to coincide with the line of the outcrop of the resistant Chemung sandstones was perhaps fortuitous, but their existence served to preserve the demarcation between the two types of drainage patterns, cuesta-making in the north, ancient tree-branch organization in the south.

Designating the Portage escarpment the dominant relief element of the Finger Lakes Region landscape ignores the circumstance that the region of the eastern Finger Lakes as a whole constitutes a broad embayment south of the Central New York lowland. From its southernmost development in the line of the axes of the Cayuga Lake and Seneca Lake valleys the Portage escarpment swings northward on both the east and the west so as to approach closely the line of outcrop of the Onondaga limestone. This embayment, it may be inferred, was due to the erosive activity in the Cayuga Valley of a river and its tributaries—the major drainage system of the area. The Cayuga "River" attained this ascendancy among north-flowing streams because the strata in its axis were less resistant than those on meridians to the east and west, especially those of the Chemung formation at the top of the pile. Seneca, Owasco, Skaneateles, and Otisco "Rivers" functioned similarly to extend the embayment southward. Although the embayment, as a whole, was not a plain, it was a lowland in relation to the upland regions surrounding it. In any event, whether due to superior erosion or inferior resistance the embayment was preglacially in existence, and its presence materially affected the further topographic evolution of the Finger Lakes Region.

The foregoing pages present in outline the sequence of development and the nature of the large units of the Finger Lakes Region relief. In general, these have significance only as features discernible on a map. Knowledge of their existence, however, is essential to understanding the land-form phenomena that present themselves to the observer in the field. In turn, it is by interpretation of the features comprehended on inspection that the origin and history of the large units emerge.

If the role of major north-flowing drainage course is assigned to a preglacial Cayuga River it is fitting to begin an account of specific topographic features with the preglacial history of that stream. It started as a short north-flowing stream on the peneplain surface developed after the first uplift of the Central New York sediments. The water parting between this stream and a corresponding stream with a southward course extended east to west approximately as a line starting near Jamesville, swinging southward to Aurora on Cayuga Lake, to Bellona near Seneca Lake, then northwestward through Bristol to a point south of Avon. This is admittedly a sketchy reconstruction but is supported by significant evidence at key points. In the Cayuga Valley the course and valleys of Big Salmon and Little Salmon creeks are relics of the south-flowing drainage that was then existent beyond, south of, the water parting

Tortuous channel of lower end of gorge, Watkins Glen, showing alternate beds of sandstone and shale composing the gorge walls. (Photograph from the New York State Department of Commerce.)

Peneplain uplands south of Newfield, New York, showing the low undulating relief of the summit tracts south of the Portage escarpment. The source area of a south-flowing stream is in the middle distance. (Photograph by O. D. von Engeln.)

Escarpments and Cuestas

in the Cayuga meridian. The separation of the two Salmon Creek streams by forking at their point of junction and the acute angle at which Big Salmon joins the Cayuga Valley together make a pattern that quite unmistakably reflects an original southward course. The acute angle of junction with the later north-directed Cayuga flow is an example of the phenomenon called "hooked tributary," a development normally interpreted as indicative of a reversal in drainage direction. The forking pattern is conspicuously manifested in the two arms of Keuka Lake, also by Canandaigua Lake and West River and obscurely by the Canadice and Hemlock lakes valleys.

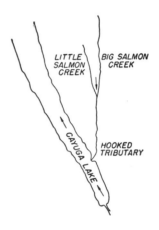

Hooked tributary, Salmon Creek. This sketch map shows the angle of junction, acute upstream, of Salmon Creek and Cayuga Lake. The course of Salmon Creek is a remainder of the preglacial southward drainage over the peneplain surface. The headwater erosion of the dominant north-flowing Cayuga "River" captured this tributary to earlier south-flowing drainage and reversed its outlet direction. The preservation of this relic of the very ancient drainage pattern is due to the action of glacial erosion reaming out and deepening the valley. Such excavation by the ice resulted because of the north-to-south trend of the Salmon Creek course.

The Salmon Creek occurrence is clearly a relic of peneplain drainage courses before the second-uplift. The three other instances cited are remainders of patterns after the second-uplift. The postpeneplain uplift apparently did not shift the line of the peneplain divide but did impart a steeper gradient to north-flowing streams. Because of this and in consequence of the belted structural pattern open to their attack, the north-flowing streams had a tremendous erosive advantage over their south-flowing competitors. To the north the conditions were perfect for a sequence of cuesta excavations; to the south the drainage could only wear through each resistant bed by slow grinding down into its top surface.

Of the north-flowing streams the one on the Cayuga meridian

was the most successful in extending its length and increasing its volume. It worked its way southward by headwater erosion to a point near Lockwood on the present Cayuta Creek. In the course of this encroachment on the south-flowing drainage it diverted first the Salmon Creek waters and then, through development of lateral tributaries in the cuesta-forming attack, in turn Fall Creek, Catatonk Creek, and the upper waters of Cayuta Creek. The rock floor of Fall Creek above the gorges is at 900 feet above sea level, of Catatonk Creek at approximately 1,000 feet. This difference in level of the tributary junctions with the main stream is indicative of the preglacial gradient of the Cayuga River.

The Fall Creek and Catatonk valleys are both broad and deep, and were evidently not at all enlarged by glacial erosion. In cross-section profile they have gently inclined dip-slope sides on the north, steep Portage-escarpment slopes on the south. On first inspection their dimensions appear extraordinarily great for lateral valleys tributary to the valley of a master north-flowing Cayuga River developed in postpeneplain time. It is, however, to be remembered that in the same interval the bold and high Niagara, Helderberg, and Onondaga escarpments had been etched into relief and wide lowlands developed at their bases, e.g., the Ontario Plain and the Mohawk Valley.

Moreover, the process of such sculpture is still in progress and can be observed in actual operation at sites where the factors at the moment (geologically speaking) are especially favorable for its visible demonstration.

On Fall Creek along N.Y.S. Route 322, just east of Forest Home village, at a point locally known as "Flat Rock" a wide expanse of rock-bottom stream bed is exposed. This is the top surface of one of the durable sandstone layers of the Finger Lakes stage (Portage) beds. It yields only slowly to the downgrinding of the stream. But the gentle inclination (dip) southward of the beds concentrates the current against the south bank, where weak shale layers overlie the resistant bed. These shale layers are obviously being progressively undercut and thus made to crumble piecemeal into the stream. The low cliff that has been developed here represents in miniature

Escarpments and Cuestas 23

the Portage escarpment, and could the course of Fall Creek be freed of its present-day surficial encumbrance of glacial debris the manifold repetition of this small-scale attack would ensure the very rapid (geologically speaking) retreat of this segment of the escarpment front. Here, quite clearly, change is being effected in a single year; tens of millions of years were available for the creation of the valley and the escarpment which bounds it on the south.

That such lateral migration of the stream's course, because of the southward dip of the beds and the presence of resistant layers in the vertical sequence of strata, was actually the preglacial history of excavation of the Fall Creek valley is impressively confirmed by the nature of the major units of topography comprising the landscape farther upstream in its course. On N.Y.S. Route 13, beyond the hill east of Varna, there is a long stretch of road that to the north overlooks a wide expanse of level country with a gentle southward slope. This obviously is the floor of the preglacial dip slope leading up to the base of the clearly defined steep and high escarpment slope on the south side of the valley along the base of which the road runs.

The retreat of the escarpment front is not, however, due solely to the downdip migration of the stream course and the lateral erosion this brings about. The face of the escarpment is furrowed by numerous gullies and at intervals by stream channels of considerable size. One such channel is the upper course of a brook that flows north down the escarpment slope parallel to Ringwood Road, which intersects N.Y.S. Route 13 about eight miles east of Ithaca. The raw sides and bottom of its channel are convincingly indicative of the notable erosive activity that takes place there at times of flood flow from heavy rains or snow melting. Collectively such streams and the minor gullies rake down the escarpment face and cause its slope to retreat, without change in steepness, parallel to its earlier positions.

Incidentally the southward dip of the rock beds directing the lateral erosion at Flat Rock, is shown in section and demonstrated to affect the whole rock pile on the east side of the Ringwood Road. Here may also be seen the alternation between shale layers and thin

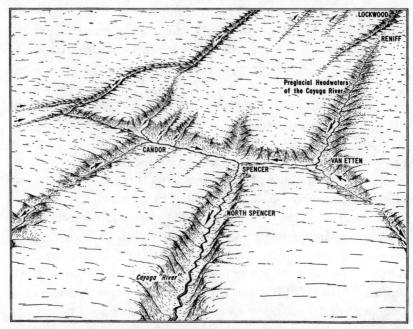

Preglacial Cayuga "River" drainage. This diagram shows a reconstruction of the inferred preglacial Cayuga drainage with the source area at Reniff and the valley floor level at Ithaca at the altitude of 900 feet. (Diagram by R. Dockendorf.)

bedded sandstones characteristic of the rocks of the upper Finger Lakes stage (Portage).

The manifestations of escarpment retreat and valley development seen along the course of Fall Creek are present on a larger scale in the Candor-Van Etten valley followed by N.Y.S. Route 223. This valley has dimensions comparable to those of Fall Creek but is no longer a major drainage line. It is a relic of the preglacial Cayuga River system rather completely deprived of its former volume of drainage by the reordering of stream patterns resulting from ice erosion and deposition. The south side of the Candor-Van Etten valley is a well-defined segment of the Portage escarpment in the line of the Cayuga Lake meridian. Although lateral erosion at its base has been terminated by the glacial diversion of drainage, the retreat of the escarpment front is still in active progress. A number of parallel streams descend its steep slope, cutting

Escarpments and Cuestas

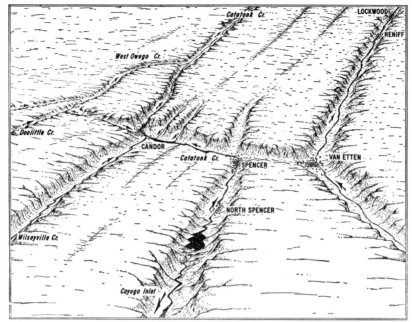

Postglacial Cayuga drainage. This diagram of the existing drainage lines of the area shows the water parting between the Susquehanna and St. Lawrence drainage in the morainic belt north of the lake near North Spencer. (Diagram by R. Dockendorf.)

deep furrows, and at their heads are eating back into the summit uplands. The relatively gentle north side of the valley is similarly occupied by parallel streams. These are the headwater remainders of south-flowing peneplain drainage cut off long preglacially by the progressive encroachment of the lateral Candor-Van Etten stream.

Despite the diversion of drainage that has left the Candor-Van Etten valley a virtually dry stream course, the process by which it was originally developed is still in operation and will soon (geologically speaking) significantly extend its length. At its east end there is a small unnamed west-flowing stream descending from a minor divide between its headwaters and those of a similar small stream (a tributary to Doolittle Creek, itself a tributary to West Owego Creek) flowing east. Since the west-flowing stream has the steeper gradient, down 650 feet compared to a descent of 450 feet only in the same distance of the tributary to Doolittle Creek, it will encroach on and progressively reverse the drainage of the tributary to Doolittle Creek until its headwaters reach the point where the Doolittle tributary joins Doolittle Creek. Then the upper waters of Doolittle Creek will abruptly be diverted into the course of the west-flowing stream, and this sudden large increase in volume will greatly increase its erosive effectiveness. The two diagram maps here presented show the present pattern of drainage and the preglacial system.

The valleys of Cascadilla Creek and of the eastern branch of Sixmile Creek and, west of the Cayuga axis, the Newfield-Pony Hollow valleys are minor counterparts of the Candor-Van Etten development. It was through the structurally conditioned erosion of these and other lateral troughs that the parallel and overlapping segmentation of the Portage escarpment was brought about.

The Seneca Lake preglacial river was the largest rival of the Cayuga River. By comparison with the Cayuga system the southward extension of the Seneca River by headwater erosion was rather limited. It reached only to Millport about ten miles south of Watkins Glen. One reason for its lesser success was the absence of any considerable tributary drainage. Further, exceptionally massive sandstone beds appear in the rock section just south of Watkins.

Other preglacial north-flowing streams that could have been rivals of the Cayuga River drainage system are those aligned with the axes of Owasco and Skaneateles lakes. These had their early southward extension terminated and their headwaters diverted by the extremely effective erosional lengthening of the east-west Fall Creek stream. As Fall Creek successively tapped the streams flowing northward to the valleys of these two lakes, it tied them into the Cayuga River system. Their northern ends were in part reversed in direction and became tributary to Fall Creek. Their beheaded southern ends, likewise now tributary to Fall Creek, continued their headward-erosion activities with undiminished volume and a steeper gradient. They were thus able to penetrate beyond the Portage escarpment front comprising the south side of the Fall

Escarpments and Cuestas

Creek valley. Thus the original headwaters of the Owasco River, as Owego Creek, worked southward to the village of Richford, 12 miles south of Dryden by way of N.Y.S. Route 38. Those of the Skaneateles River worked back to Messengerville on the Tioughnioga River, south of Cortland on U.S. Route 11.

Of the Finger Lakes west of Seneca Lake, the valleys of Canandaigua, Honeoye, and Conesus preglacially had short north-flowing streams with divides either within the length of the present lake basin or only a short distance south of its head. Both arms of Keuka Lake are definitely in valleys of preglacially south-flowing drainage. Hemlock and Canadice lakes appear also to be of this pattern. In effect all the Finger Lakes west of Seneca Lake have basins that either straddle preglacial divide sites or are entirely aligned with and superposed on the headwater valleys of south-flowing preglacial peneplain streams. This circumstance is the quite sufficient warrant for making a major distinction between the Finger Lakes "East" and the Finger Lakes "West." This separation of the lakes into two groups with dissimilar backgrounds is a significant contribution to the comprehension of the land-form evolution of the region.

Summary

The guiding factor in the development of the land forms of the Central New York lowland (of which the Finger Lakes Region is the southern embayment) is the rock structure. This consists of a vast pile, 8,000 feet thick, of beds of limestones, sandstones, and shales gently inclined (dipping) southward. After an initial uplift of unknown height the top surface of the pile was worn down by rain and rivers to a near-sea-level plain of low relief (peneplain). In consequence of this leveling, the northern edge of the beds was shaved, so that, because of the dip, the deeper, older beds were exposed as a series of bands extending east-west across northern New York and southern Ontario.

These beds varied greatly in their resistance to weathering (crumbling, chiefly through solution and frost) and erosion (the

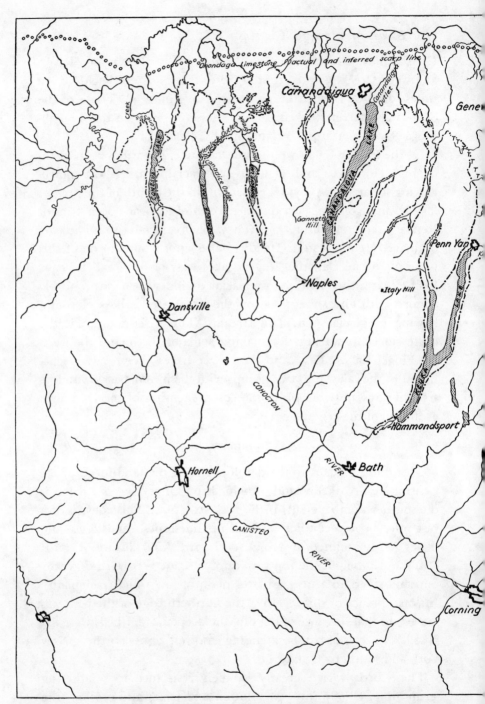

Finger Lakes "West." The outstanding characteristic of this map is the patterning of the lake basins to conform to the interpretation that they are excavations of the valleys of preglacially south-flowing streams. This is most perfectly exemplified by Keuka Lake. One arm of the matching pattern of Canandaigua Lake is dry but the erosional development is the same. Canadice and Hemlock lakes repeat the arrangement though only obscurely except as the other examples indicate it.

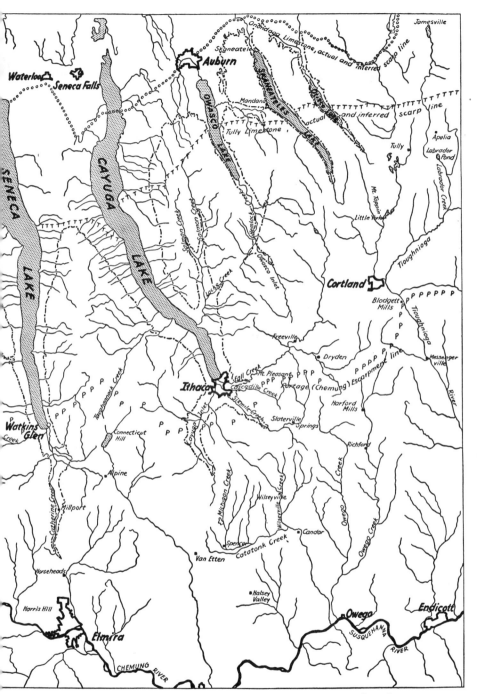

Finger Lakes "East." The "O" line marks the position of the Onondaga limestone escarpment, the "T" line that of the Tully limestone, the "P" line the general position of the Portage escarpment front. The short postglacial and interglacial consequent streams down the oversteepened slopes of the Cayuga and Seneca basins are clearly in evidence. The dot-and-dash line outlines the shores of proglacial Lake Newberry, that overflowed at Horseheads at the time of its wide expansion. The over-all dendritic pattern of the south-flowing drainage, beyond the Portage escarpment, is indicated by the courses of numerous streams.

wear of rivers). Most resistant were three massive limestone formations, the Lockport (Niagara), Helderberg-Onondaga, and Tully, in that order as they are encountered going from north to south. Still farther south and higher in the pile, sandstone beds, primarily of "Portage" and Chemung age were similarly resistant, but much more variably so than the continuous thick limestone layers. Between the resistant beds were very thick (relative to the limestone strata) beds of shale nearly all of which were exceptionally susceptible to crumbling and could quickly be worn away by stream action.

The peneplanation was succeeded by another uplift which did not significantly disturb the arrangement or dip of the beds. This uplift occurred about thirty million years ago. Thereafter until the Pleistocene glacial epoch, about one million years ago, a renewed wearing down was in progress.

Although low, the relief of the peneplain was sufficiently great for streams to flow over its surface to the sea. Moreover, there was probably a water parting extending in a line southwestward from the southwest base of the Adirondacks. From this divide streams wandered northward toward what is now Hudson Bay and southeastward to the Atlantic. No evidence remains of the pattern of the northward drainage beyond the Central New York lowland, but the courses of the Susquehanna and the Delaware are in the main survivals of the trunk peneplain drainage southeastward. The disappearance of the major northward peneplain streams is due first to the pattern of the beveled durable and weak formations they crossed and second to the relief modifications resulting from the ice invasion of the Pleistocene glacial epoch.

As detailed above, the north-flowing streams in crossing the alternating strong and weak east-west rock belts initiated the development of cuesta topography. In brief, cuestas are ridges with one steep slope, the escarpment slope, and the other gentle, the dip slope. These forms resulted here because east-west tributaries to the major northward trunk streams were extremely effective in excavating the weak belts of rocks and eventually creating lowlands over the dip-slope surface of the underlying durable formation.

Of the several cuestas the one fixed by the line of outcrop of the Niagara (Lockport) limestone is most prominent. The Onondaga cuesta becomes obscure north of the Finger Lakes, but its line marks the north end of the embayment of the Central New York lowland, which is terminated in the south by the irregular and discontinuous Portage (Chemung) escarpment. This embayment has resulted from the effective headwater erosion of weak beds by the stream systems of the two north-flowing trunk rivers located on the axes of the present Cayuga and Seneca lakes valleys.

In brief, the successive escarpments that emerged as the result of the postpeneplain dissection of the area are the basic elements of Central New York topography. Although other features are scenically more conspicuous and attention-arresting, the cuesta form remains as the structural background on or against which the lesser phenomena are positioned.

2 · *The Peneplain Uplands and Their Dissection*

SOUTH of the irregular line of the Portage escarpment are the peneplain uplands. Specifically these are the essentially level summit areas that stand at the approximate altitude of the escarpment front, 1,600 to 1,800 feet. Unit tracts range from a few acres to four square miles in extent.

That the uplands are the remainders of an earlier peneplain surface is deducible from several lines of evidence. In the territory within sight from any one of the upland tracts there is a rather perfect accordance of summit level. In view of the uniform southerly regional dip of the strata this means that the edges of the beds have been truncated to a common level and that as one goes southward beds higher and higher in the pile will crop out at the common upland level. In other words, the presence of summit tracts at the same altitude is in the over-all sense not a reflection of structure. That is, the high points do not regularly coincide with the crests of the upbends of the rock beds. Neither are the uplands here, as is commonly true in the plateau regions of the American West, structural

The Peneplain Uplands

plains—that is, level surfaces of large or restricted extent composed of the top surface of a single especially resistant flat-lying bed, usually a massive sandstone.

It is not, however, permissible to assert that the upland summit surfaces of the Central New York region in no wise reflect structure. They do possess a slight measure of relief—up to several hundred feet at particular points. Since the advent of television this moderately greater altitude has led to the searching out of such spots as sites for broadcasting and signaling towers. Singularly enough, the presence of these tracts of superior height is significant evidence of the peneplain origin of the upland surfaces, as the high points coincide with synclinal axes (troughs) of the sedimentary pile.

In the preceding discussions of the rock structure of the Finger Lakes Region it was implied that there was no departure from a uniform slight regional dip southward. Although such is, broadly considered, the case, it is not a finally exact statement. The lateral thrust which bent the strata of the Appalachian Mountain region of Pennsylvania into great undulating upfolds and downfolds was not wholly ineffective in the Central New York area. Although the major response to the force there was the near-vertical raising of the sea floor, with its accumulated pile of sediments, some slight remainder of the lateral thrust was transmitted beyond the mountains into the area of plateau uplift farther north. This sufficed to introduce a succession of low broad ripples, with crests and troughs parallel to the Appalachian Mountain folds, into the Central New York strata.

If this original structural disturbance were topographically preserved, then the upbends (anticlines) should appear as the high places of the upland surface. But the reverse is the case. The downbends (synclines) are now the highest summit tracts. This is the logical outcome of the wearing down of folded rock structures to peneplain surfaces by rain and rivers. The upbends are attacked first and most vigorously, and by a complicated sequence of drainage adjustments the end result in the peneplain stage is that the synclinal belts are the last to be reduced in the ultimate leveling down.

That this development is apparent with so faint expression of fold structure as appears in the Central New York strata is indicative of the degree of completeness of the peneplanation achieved. It is well exemplified by Connecticut Hill, southwest of Ithaca, the highest point, 2,095 feet, in this general area. The stratigraphy of a section through this summit has been carefully mapped, as shown in the accompanying diagram. The perfect coincidence of the Connecticut Hill summit and the synclinal axis is quite decisive for establishing the peneplain history of the surface. The average altitude of the uplands in the neighborhood of the Portage escarpment belt is 1,800 feet. The summits along the synclinal axes average 100 to 200 feet higher.

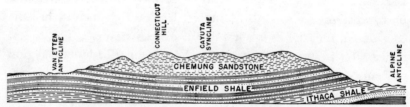

Connecticut Hill. This structure diagram through the hill illustrates how the highest elevations of the peneplain upland surface occur where the rock strata are downbent, have synclinal structure. Such discordance between altitude and structure is evidence of the peneplain origin of the upland country. (From Folio 169, Watkins Glen–Catatonk, U.S. Geological Survey.)

The upland tracts are widest and most continuous close to the escarpment crest. Because the Portage escarpment line is so irregular, isolated areas of upland appear north of its general trend and separated from this average line by broad valleys. In part, these isolated summits are due to the reinforcement of the Portage cliff by unreduced remnants of the lowland which should have developed over the Tully dip slope. The territory north of the East Branch of the Tioughnioga River between Truxton and Apulia (N.Y.S. Route 91) exemplifies this development. Once past the zone where the structural factors are thus intermingled a clear-cut and

The Peneplain Uplands

distinctive pattern of drainage is evident. The uplands are dissected by a multitude of small streams which have their minute headwater ends near the escarpment crest. These in turn are tributary to trunk streams which join the Susquehanna River. The Chenango and Otselic rivers are representative of such main stems. Others do not qualify so well as typical instances because their valleys have been too greatly modified and their courses reordered by glacial erosion and deposition.

But in the sections where the postpeneplain drainage development south of the line of the Portage escarpment has been best preserved the dendritic pattern is quite diagrammatically present. This is in accord with the expectable sequence of structure, process, and gradient as postulated. The perfection of feathering out to headwater brooks is greatest in the east and west limbs of the central embayment of the lowland. This ultimate fineness of subdivision in the source areas of the trunk streams indicates that these headwaters mark the line of the divide before the uplift of the peneplain. In the central embayment the potent north-flowing streams in the axes of the Cayuga and Seneca valleys had in postuplift time extended their drainage basins so far southward by headwater erosion as to reverse the flow of much of the water in their source region near the divide.

If it were assumed that structure exercises no control over the pattern of valley development by river erosion, then each stream system would comprise a trunk in a straight line down the general slope of the land with straight-line tributaries joining it from the right and left and these in turn fed by smaller branches until finally the twigs represented by the headwater brooks are reached. This is completely the opposite of the cuesta sequence in operation in the postuplift erosional history of the territory north of the peneplain divide.

In a wholly unguided development, as conceived above, the tributaries would come in like the branches on an evergreen tree, and the system would have a pointed apex, like that of a pine or spruce. However, although massive resistant layers to promote cuesta emergence are absent in the upper "Portage" and Chemung strata, there is enough alternation in the durability of the beds to introduce some degree of structural guidance. In consequence, the tributaries to the south-flowing trunk streams tend to curve around in their lower courses so that the dendritic pattern resembles that of an elm or an oak rather than that of a needle-leaf evergreen. It also resolves into a bushy instead of a pointed top. The main Canasawacta Creek and its East Branch, both tributary to the Chenango River at Norwich, and their small tributaries and subtributaries are an excellent example of the dendritic pattern and of this effect. The more resistant layers do tend to emerge as minor escarpment slopes on the south sides of the tributary streams, whereas the north sides of their valleys are characteristically gentle dip slopes.

Since, however, the courses are uniformly downdip, stream erosion must of necessity be primarily grinding through the beds from their top surface, and not undermining. Accordingly, waterfalls and cascades are lacking. Although the major waterfalls that decorate the north-flowing drainage are occurrences chiefly due to glacial action, small north-flowing streams that flow on rock floors commonly cascade over outcropping ledges of slightly superior resistance to erosion. The complete absence of even such minor departures from a smooth gradient is perhaps the outstanding characteristic of difference between the drainage north of the Portage escarpment and that of the peneplain upland country south of it. What is, perhaps, even more remarkable is that, although the peneplain upland province was subject to the same massive glacial invasion as the cuesta belt on the north, no similar gorge and waterfall phenomena resulted. Evidently the dip-slope structural situation inhibited glacial processes as well as stream erosion.

Another circumstance indicative of the preservation of the stream system as it originated on the uplifted peneplain surface is that tributary streams all join main streams with the acute angle of junction upstream. This is the normal pattern for drainage systems free from outside interference during the course of their develop-

Keuka (or Crooked) Lake, the type example of the Finger Lakes "West." The two arms of the lake represent the branching headwater valleys of south-flowing streams excavated by glacial erosion into rock basins for the lake. (Photograph by C. S. Robinson.)

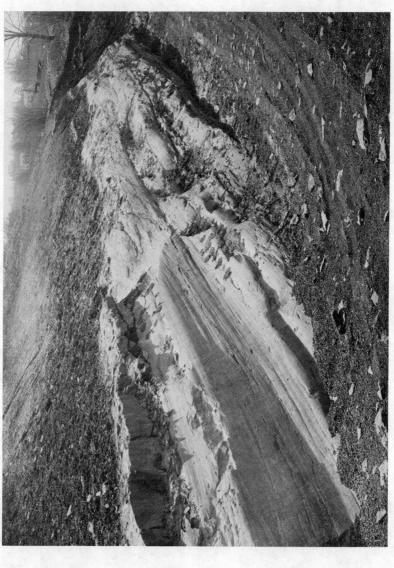

Glacial erosion by grinding. This rock ledge on the east side of the Cayuga Lake valley, north of Ithaca, was scoured and grooved by the Cayuga lobe of the continental glacier. Perfectly preserved for 9,000 to 10,000 years because it was buried under ground moraine, it was uncovered in course of excavation for building. (Photograph by G. F. Morgan.)

The Peneplain Uplands

ment and not subject to structural guidance. Had the trunk streams experienced reversal of flow-direction through captures or land tilting, evidence of such interference would appear in the phenomenon of hooked tributaries, that is, tributaries whose angle of junction in relation to the main stream is obtuse upstream.

Finally, the longitudinal profiles of the valley floors of these dendritic-pattern streams exactly conform to the ideal curve for such a history of valley erosion. This is the more remarkable because it might be expected that glacial deposits could be so thick in places as to interrupt its continuity. But, as illustrated by the tributary headwaters of the East Branch of the Canasawacta Creek in the Chenango River system, the curve is steep, nearly vertical, at the source and concave downstream at an ever-decreasing gradient.

In brief, it appears that the landscape of the peneplain uplands and their valleys is distinctly unlike that of the strongly lineamented configuration of the cuesta lands of the lowland areas to the north. It has instead a regularity and repetition of form that bespeaks a uniform evolution over a long period. It did not, however, wholly escape modification by glacial action.

3 · *Finger Lakes "East"— Finger Lakes "West"*

THE features that warrant the claim of uniqueness for the Finger Lakes Region derive primarily from the invasion and long occupation of a landscape of its particular configuration by the continental glaciers of the Pleistocene epoch. Nowhere else in the world, so far as I can learn, did the ice advance against and override escarpment ramparts extending directly athwart the direction of glacier flow.

Years ago I visited the north front of the Harz Mountains in Germany because the writings of German geologists intimated that the European continental glaciers had there encountered and entered upon a barrier region of preglacial topography quite similar to that of Central New York. It was, however, immediately evident on field examination that, while the glacier margin had advanced almost to the base of the Harz Mountain front, the upland territory itself had not been invaded. The highlands exhibited all the land-form features appropriate to an uninterrupted history of weathering and stream erosion. I know of no other place within the regions covered by the North American or European continen-

Finger Lakes "East" and "West"

tal glaciers that presented to the invading ice a topographic configuration like that of Central New York. This, accordingly, is one aspect of the uniqueness of the Finger Lakes Region.

By itself, however, such topographic confrontation of the advancing ice would not effectively sustain an assertion of notable uniqueness. This claim is more specifically dependent on the particular circumstance that the south-extending embayment of the Central New York lowland acted as a funnel to crowd the ice into broadly open north-sloping valleys of preglacial erosion. This crowding was most pronounced in the two great axial valleys of the embayment, those of the Cayuga (lake) and Seneca (lake) rivers. These two major streams had worked their way back by headwater erosion beyond the front line of the Portage escarpment, and, indeed, across that of the divide line on the peneplain surface. Farther east similar north-flowing streams of smaller size, whose valleys are now occupied by Owasco, Skaneateles, and Otisco lakes, also provided open-arm invitation to the advancing ice, and these lake basins are the counterparts on a reduced scale of the Cayuga and Seneca basins. Collectively these five lakes are here termed the Finger Lakes "East." (The Onondaga Valley extending south from Syracuse, while not occupied by a lake, on the basis of origin and characteristics also belongs in this category.)

A north-south line halfway between Seneca Lake and Keuka Lake separates the province of the Finger Lakes "East" from that of the Finger Lakes "West." Keuka Lake, the farthest east of the western group is also locally known as Crooked Lake, and this colloquial appellation is significant because it directs attention to the characteristics by which the difference between the two groups of lakes, east and west, is most obviously indicated—the branching Y form of the western group with the broad stem of the Y directed to the south.

Keuka Lake is at once the most striking, completely developed, and representative example of the western group. The others are, from east to west, Canandaigua, Canadice, and Hemlock. (These listings omit Honeoye and Conesus lakes, because they conform in origin and pattern to the eastern group.) The perfection of Finger Lakes "West" form shown by Keuka Lake appears to be owing primarily to its position farther south than any of the others of the western group. This position ensures its being a reflection, relic, or inheritance of the pattern of the preglacial drainage south of the peneplain divide. Canandaigua Lake has a similar tributary valley (not lake occupied), with two arms preserving the earlier pattern of drainage, and the alignment of Hemlock and Canadice lakes is suggestive of the same guidance. In other words, Keuka and other (Canandaigua, Hemlock, and Canadice) Finger Lakes "West" occupy basins which are modifications of the headwater branching valleys of the south-flowing peneplain drainage. When the glacier ice entered these valleys, it proceeded downgrade, whereas in the Finger Lakes "East" it went upgrade.

The creation of the lake basins of the Finger Lakes "West" by down-valley glacial erosion is not a unique development. The fiords of Norway are a completely analogous phenomenon. The continental ice sheet there overtopped the divide of the Kjölen Mountains and excavated the fiord basins on their far side. The Swiss-Italian lakes on the south side of the Alps, the lakes of England's Lake District, and, nearer home, the Saguenay Valley and the Hudson Valley are other instances of down-valley ice erosion.

Further evidence of the validity of the distinction here made between the two groups of lakes is that the Finger Lakes "East" converge northward whereas those of the Finger Lakes "West" converge southward. Otherwise stated, the line of the preglacial peneplain divide had been preglacially shifted far southward by headwater erosion of the Portage escarpment in the region of the Finger Lakes "East." The Salmon Creek valley, which joins the Cayuga Lake valley a few miles north of Ithaca, is a remarkable relic of the drainage that existed before the reversal took place. It is a hooked tributary with the acute angle of junction on the downstream direction of the main stream. In origin its valley apparently harks back to the time when this reach of the Cayuga Lake valley

Finger Lakes "East" and "West"

was occupied by a south-flowing stream. Thus Salmon Creek is interpreted as a relic of the predissection postpeneplain drainage.

At other places within the Finger Lakes "East" province, where the preglacial southward encroachment of the north-flowing streams had not progressed so far as in the axial Cayuga-Seneca section, quite revealing, sharply defined, preglacial divides existed where the heads of north-flowing and south-flowing streams were directly opposed. These divides were effectively breached, by and in consequence of glacial action, but their sites are clearly marked by the narrowing of the valleys from both directions to a defile where the uplands on each side of the valley attain their highest altitudes.

This phenomenon has quite diagrammatic expression between Richford and Harford on N.Y.S. Route 38. The preglacial north-flowing stream went via Dryden to the valley of Fall Creek, the south-flowing one to join the Susquehanna at Owego. A similar development occurs in the Tioughnioga River valley three miles north of Messengerville on U.S. Route 11 between Cortland and Binghamton. In both instances the drainage is now southward across these former divide sites; this is because the valley floors were so deeply aggraded from the north by gravel and sand fills at the close of the glacial epoch as to provide a continuous south gradient for postglacial drainage. But examination of the topographic maps of the divide areas reveals that tributary streams from the uplands still retain the directions appropriate to each side of the divide where not diverted by glacial deposits. An east-west line connecting these divide sites serves to give the approximate position of the water parting just before glaciation altered it.

There is a possibility of confusion in regard to the divide lines referred to above and in the preceding chapter. It is postulated that following the uplift of the peneplain the divide on the peneplain surface continued to function in directing drainage to the north and south. It is maintained, further, that the general position of this line is still preserved in the headwater areas of the Chenango and Otselic rivers and, farther east, with reference to the Susquehanna drainage, because these streams feather out in their source regions into a pattern of fine tributary feeders and there is no evidence of a trunk stream transgressing this line. The same interpretation applies to the province of the Finger Lakes "West." There, however, the fine detail of the source-area streams has either been almost completely erased by glacial erosion or obscured by glacial deposition.

These relics of the peneplain divide were functioning preglacially and continued to do so postglacially. But in the axial section of the embayment, especially, and in lesser measure in the rest of the Finger Lakes "East" region, the *preglacial* divide had been displaced far south of the *peneplain* divide by the invasion of Fall Creek, tributary to the Cayuga River. The breached divide sites at Harford (N.Y.S. Route 38) and at Blodgett Mills (U.S. Route 11) clearly mark points where the preglacial divides had been shifted by headwater erosion of reversed south-flowing streams (because of the eastward erosion of Fall Creek) and the preglacial divide established on a line considerably south of an earlier preglacial but postpeneplain divide.

In the axial section marked by the Cayuga and Seneca valleys the preglacial divide had been moved so far south that evidence of the line of the previous water parting is almost completely lacking. The source of Salmon Creek, the hooked tributary to the Cayuga valley, is about as definite a clue to its position as still exists. South of Seneca Lake the preglacial divide was in the Millport–Pine Valley reach of N.Y.S. Route 14; south of Cayuga Lake it had been pushed back to the vicinity of Lockwood on N.Y.S. Route 34. To the east of this line, in Halsey Valley on an unnumbered road south from N.Y.S. Route 96, there is another breached divide like those at Harford and Blodgett Mills, but on a smaller scale—perhaps, therefore, a more convincing example of the phenomenon. This case is interesting, moreover, because its position coincides with that assigned to the Lockwood reach. The hooked tributaries of the reversed stream, South Branch, flowing down to Catatonk Creek are further evidence of the preglacial shifting of divides in Halsey Valley.

Considered by itself, perhaps no one of the kinds of evidence set forth above would be accepted as finally convincing in regard to the differentiation here proposed between Finger Lakes "East" and Finger Lakes "West." But taken into account collectively, the case for the distinction becomes very strong. And if this is regarded as demonstrated, interest in the over-all phenomena of the Finger Lakes and their scenic features should be considerably enhanced, because variations in aspect on one side or the other of the dividing line may then be correlated with their different histories of origin and development.

In the area of the Finger Lakes "East" the preglacial pattern of stream courses in the upper reaches of the Fall Creek valley is both pertinent to such an interpretation and an elegant exemplification of the orderly sequence governing the establishment of the pattern.

In preceding pages the erosive competence of streams flowing parallel to the base of escarpment slopes was described. The extension eastward of the Fall Creek tributary to the Cayuga River was cited, and the nature of the valley that resulted from such action was set forth. But for an example of the lengthening, in progress, of such a valley recourse was had to the topographic situation in the Candor-Van Etten valley, which, like the Fall Creek valley, was developed parallel to the base of an escarpment slope. At the Candor location the process of headwater growth in valley length is clearly in evidence.

Fall Creek, however, does not serve to illustrate this phenomenon of extension in length because its present source is not at the preglacial site. Preglacially Fall Creek had its origin at the head of what is now the East Branch of the Tioughnioga River. The reach between where Fall Creek now enters the Fall Creek valley and where the course of the East Branch of the Tioughnioga River is encountered—the area occupied by the city of Cortland—was built up so high by a vast deposit of glacial gravel that the preglacial Tioughnioga flow was prevented, postglacially, from continuing its earlier course westward down the Fall Creek valley.

As with the other Finger Lakes "East," the preglacial streams in the axes of the present Owasco, Skaneateles, and Otisco lakes and the Onondaga Valley were north-flowing and had extended their courses southward by headwater erosion beyond the line of the present Fall Creek valley. At their sources they were encroaching on the ancient southward dendritic drainage. Meanwhile the persistent extension eastward of Fall Creek successively brought about the capture of these streams. This may seem a tremendous erosive task, but time on the order of 30 million years was available for its accomplishment.

When a north-flowing stream was thus intercepted notable readjustments of stream flow immediately ensued. Since the valley floor of the invading Fall Creek was appreciably lower than the level of the channel bottom of the intercepted north-flowing stream, its upper waters from the south at once, in effect, cascaded into Fall Creek. The beheaded continuation of the north-flowing stream on the other side of the Fall Creek valley flowed on as before but with diminished volume. A short reach on the north side of the Fall Creek valley had its course reversed to a southward direction, and since this reversed course had a steep descent it shortly developed into a deep gully which, despite a limited drainage volume, eventually was so enlarged and extended that its headwaters reached a new divide situated a considerable distance north of the Fall Creek valley side. Here again it must be repeated that these were not adjustments made in a hundred or a thousand years, but were shifts achieved in the course of millions of years.

Eastward from the Cayuga Valley the first north-flowing stream to be diverted by the insidious headwater encroachment of Fall Creek was one aligned with the axis of the Owasco Lake valley. The relic headwaters of the earlier north-flowing stream in this instance is the short course of Dryden Creek. The earlier divide, however, was some miles to the south at Richford. The valley reach between

Finger Lakes "East" and "West"

Richford and Harford was so deeply aggraded by glacial outwash gravels as to build up a southward gradient for Owego Creek across the divide site.

Along the axis of the Skaneateles Lake valley the same voluminous aggrading, in this instance of the Tioughnioga River valley, extends across the divide site north of Messengerville so that no relic of the earlier north-flowing drainage remains. But in this case the development of its reversed south-flowing section is clearly in evidence as Factory Brook. This starts a few miles south of the village of Scott, where what may be called the secondary divide was established subsequent to the bisection by Fall Creek. Incidentally, the present south-flowing headwaters of Fall Creek starting in Lake Como may be another example of a reversed drainage course resulting from intersection, but the relations are not clearly discernible.

In view of the simple structure of the rocks of the Central New York area and their history of only two uplifts, and those vertical without significant disturbance of the attitude of the strata, interpretation of the valley history of the Finger Lakes Region would be easy except for the intervention of the glacial invasions. Through erosion and deposit the ice so modified the preglacial landscape that its reconstruction presents difficulties. That the glacial occupation did not completely obscure the picture indicates the basic role of the rock structure in establishing and maintaining the characteristic major elements of the relief of the region.

4 · *Glacial Erosion and Rock-Basin Lakes*

UP to this point references to glacial action in the Finger Lakes Region have, for the most part, been general. Since, however, the major scenic element of the area is the presence of the lakes and since these and the associated spectacular features, the waterfalls and gorges, result from the glacial invasions of the region, it is essential for an appreciative understanding of the phenomena that an analysis of their origin through glacial action be presented. The determinant process in the creation of these striking landscape forms was glacial erosion.

Stream erosion, the carving of valleys by running water is open to observation and experiment. The action of current-transported sand and pebbles in grinding down the rock bed over which the water flows is obvious. The violent disruption of the rock on the bottom and sides of a narrow valley subject to a flood becomes clear as soon as the stream subsides to normal volume.

It is quite otherwise with glaciers. What goes on at the bottom of the ice is not open to inspection. Experimental investigation is not possible because the minimal thickness of, say, 500 feet of ice

Glacial Erosion

of a freely flowing glacier far transcends any feasible laboratory duplication. Accordingly, study of the manner and measure of erosion by ice can only be made through observation and interpretation of phenomena present after the glacier has melted.

Because of this necessity to depend on inference and deduction, a controversy that dragged out over decades developed between geologists who were convinced of the efficacy of glacial erosion and those who maintained that the ice was quite incapable of significant rock removal. By "efficacy" is meant both the general reduction in level of the rock surfaces over which the glacier passed and differential erosion, a concentration of attack along particular lines which resulted in a heightening and reconstruction of the preglacial relief.

That glaciers were capable of scratching and abrading the rock floors and sides of valleys was generally conceded. But the waste from such grinding was held to be small in amount. In fact it was contended that, compared with the attack and destruction that could be wrought under the air by rain and rivers, the ice cover actually exercised a protective effect. Accordingly, although it was admitted that the forms of valleys and hill summits might be slightly modified by ice scour, their major sculpture was held to be due to subaerial weathering and stream erosion.[1]

It was interesting to note in going over some of the literature on

---

[1] While it is true that quantitatively scour is the minor process of glacial erosion, the visible effects of such action by continental glaciation, where preserved under a protective cover of impervious clayey soil materials, are impressive.

(There is a flat rock shelf across the street from the Boldt Hall Tower of Cornell University and a short distance up the slope that is wonderfully striated over a wide area. Part of this was uncovered years ago when some construction was in progress. A considerable expanse of this hallmark of the Pleistocene glacial epoch could be excavated, washed clean, surrounded by a low wall, and coated with a transparent plastic as a protection against frost attack, and so preserved as a natural history exhibit of outstanding interest. Today, when a multiplicity of liquid transparent plastics capable of hardening into tough impervious shells are available, it would be possible to ensure the permanency of such an exhibit.)

the subject to find my old chief, Professor R. S. Tarr, wavering in 1904 between the glacial-erosion and the anti-glacial-erosion camps, although he had previously signed up with the erosionists. In the 1904 paper entitled "Hanging Valleys in the Finger Lakes Region of Central New York" he found himself compelled to rehearse the antierosionist arguments and to have misgivings about the erosionist stand primarily because a colleague, A. C. Gill, a mineralogist, for whom he (and I) had great respect, was firmly of the antierosionist conviction—so Tarr states in a footnote. But in another paper two years later entitled "Watkins Glen and Other Gorges of the Finger Lakes Region of Central New York," Tarr places himself squarely on the erosionist side. It may be added that Professor H. L. Fairchild, of the University of Rochester, was the leading antierosionist in these parts, and he died, in 1943, without conceding that the major landscape features of the Finger Lakes Region were anything other than the product of weathering and stream erosion.

The chief difficulty in the way of earlier acceptance of a theory of efficacious glacial erosion was the failure to understand the true nature of a glacier. The earlier students of glaciology conceived glaciers to be rigid crystalline masses of ice. It was known that they moved differentially, in the case of valley glaciers, faster in the middle than at the sides, but this was not regarded as a true flow motion. Instead it was held that the movement of the glacier as a whole represented the sum of an indefinitely great number of minute breaks, fault shiftings, through the rigid ice crystals. Elaborate experiments were made on existing glaciers to show that their forward movements were spasmodic and not continuous.

Such spasmodic motion was demonstrated at the points where the tests were conducted. But these results were not valid, because the experiments were made either in the thin terminal parts of the ice or on the upper surface of the main body of the glacier. At such sites glaciers are crevassed to depths as great as 200 feet.

What was not appreciated when the minute faulting and rehealing theory of glacier motion was generally accepted is that the

Glacial Erosion

upper crust and thin terminal parts of a glacier are rigid ice carried and shoved along by the interior of viscously and continuously flowing ice. An active glacier in reality exists as an entity with a nicely adjusted equilibrium. It has been shown that the ice of the deep interior and bottom of a glacier is at the pressure-temperature melting point. The thicker the glacier the lower the temperature. But even in very large and deep glaciers this temperature is only a degree or so below the melting point of ice in air. Any variation in pressure, accordingly, tends either to promote liquidity (greater pressure) or refreezing (less pressure). The basic reason for this is that when water is converted to ice it expands; hence water pipes burst. Contrariwise, if ice is put under pressure it tends to change back into water.

This account of the physics of the interior ice of glaciers is incomplete but will suffice to make clear that instead of being a rigid unyielding mass a glacier is actually extremely mobile. Its ice can be and is forced into even minute fissures in rocks, and it flows in interior currents, if not so nimbly, yet similarly to the currents of a deep turbulent stream. When, accordingly, a glacier advances across a region of marked topographic relief, its interior flow lines are affected and directed by every inequality, hill and valley, of the preglacial landscape.

As has already been suggested, the configuration of the Finger Lakes Region was peculiarly adapted to influence such ice movements. While the general direction of the ice advance was southward, every valley and barrier served to diversify and complicate the interior currents of the glacier. The funnel form of the north-opening valleys of the Finger Lakes "East" province was ideally shaped to concentrate and direct the advance of the ice. The progressive narrowing of the valleys southward and the higher valley walls in the headwaters region of and beyond the Portage escarpment compelled a faster flow through the constricted upper courses in order that the volume of ice thrust into the funnel head would be moved through its spout. This faster flow enhanced the rate of glacier erosion of the valley bottom so that the channel was differentially deepened. In other words, the cross section area was enlarged by deepening the valley cut until it sufficed to convey the ice in the same volume as when it entered the funnel head. These circumstances applied during the period that the advancing ice was projected southward as tongues extending up the valleys.

When the ice flood had so increased in depth and volume that the whole countryside was covered and the highest hills were under a thousand or more feet of ice, the concentration of erosion in the upper reaches of the north-south valleys was continued because of the internal flow of the ice. The overriding glacier, as a whole, maintained an essentially uniform surface gradient toward its terminal front. This is clear from evidence preserved on mountain sides near the ultimate farthest advance of the ice. It can also be seen on the existing Malaspina and Hardanger glaciers. The direct effect of the differential erosion resulting from the deeper ice currents following the lines of the north-south valleys would be to induce a sag along these lines in the surface of the ice sheet. But as rapidly as the deepening progressed, any such tendency toward a surface sag was compensated for by lateral inflow of adjoining ice in sufficient volume to maintain the uniform surface slope of the glacier.

The adjustment to topographic irregularities by ice flow and currents within the glacier is a factor that has been too little appreciated in studies of the effects produced by the subjection of a region to overriding by glaciers of continental dimensions. The interior ice appears to have been a veritable maze of currents when crossing a landscape of such diverse preglacial relief as was that of the Finger Lakes Region. The currents even moved upward in the ice as is attested by finding on South Hill in Ithaca at an altitude of 1,000 feet fragments of Oriskany sandstone, an extremely distinctive and readily identifiable formation, which crops out at an altitude of 400 feet near the north end of Cayuga Lake. Only internal upward ice movement could have brought them there. In fact the evidence is clear that every inequality of land surface, original or produced by the glacier itself, served to direct deep interior cur-

Glacial Erosion

rents of ice independent of the general forward movement of the glacier.

Such were the conditions of the ice invasion of the area of the Finger Lakes "East." The question remains by what specific processes was the erosion accomplished. As was pointed out in a preceding paragraph, what knowledge of these there is rests on inference from the effects produced. The grinding done by the rock- and grit-shod bottom ice is made evident by the scratched and striated rock surfaces commonly seen when the protective cover of soil overlying the bed rock is newly removed. While not inconsiderable, the amount of rock removal due to such abrasion would not suffice to account for the scale of the glacier-induced relief changes.

There is, however, another much more effective process of glacier erosion whose nature is also clearly indicated by the effects produced. This process is called plucking, but could also be referred to as quarrying, or wedging, or prying. It operates conspicuously where rocks are massive and are cut by fine, deep cracks (or joint planes) occurring at horizontal intervals of from one to ten or more feet. The granitic Thousand Islands in the St. Lawrence River are many of them perfect exhibits of the plucking phenomenon. On the side (down-river) against which the ice impinged, grinding produced perfectly smoothed upcurved slopes. The upstream side, where the ice pulled away from the rock, is regularly a steep cliff face with a hackly surface resulting from the removal intact of blocks outlined by the joint planes. A detailed exposition of why and how this process is inferred to operate is outside the scope of this book. It will suffice here to say that the ice adheres by freezing to the rock face and that this bond is sufficiently strong to permit pulling out, as the glacier moves past, the blocks outlined by the joint cracks.

Although the rocks of the Finger Lakes Region are abundantly intersected by joint-plane fissures, the major part of the pile of strata is comprised of shales and thin-bedded sandstones of too slight coherence to yield in large units. This characteristic, however, would not preclude the plucking process from operating.

Smaller fragments could be detached in the same manner. But, because these were relatively fragile, they were quickly reduced to small unrecognizable pieces or indeed to grit and powder in the "mill" at the ice under surface. This interpretation is quite definitely authenticated by the circumstance that the two really massive formations of the region, the Onondaga and Tully limestones, are abundantly represented by large blocks torn loose from where they crop out, carried forward, and deposited at points miles to the south of the site from which they were quarried. In other words, where the rock was structurally suitable for removal in large blocks, evidence of the plucking process is strikingly present. Such fragments are conspicuous in the glacial deposits at two places—over and south of the outcrop line of the Onondaga limestone on N.Y.S. Route 173 west from Jamesville, and over and south of the outcrop line of the Tully limestone near the crest of the hill on the road off N.Y.S. Route 80, north from Apulia. In each instance big blocks of the respective formations appear at the surface in fields, in stone fences, and in road cuts.

Here, accordingly, is a process entirely competent to perform the prodigious volume of rock excavation necessary if the impressive erosional phenomena of the region are to be attributed to glacial action. Greatest of these feats is the creation of the rock basins of the Finger Lakes, two of which have bottoms (determined by sounding) below sea level. And these are not their real bottoms because glacial and postglacial sediments of unknown thickness cover the true rock bottoms. A very detailed sounding and mapping of the Cayuga Lake bottom by the Civil Engineering College of Cornell University revealed that the floor of the lake from side to side and nearly so from end to end is almost as level and featureless as the surface of an ice sheet over a mill pond. (This map exists only in manuscript in the files of the College of Civil Engineering.) The lakes are deepest at their narrowest parts somewhere about the middle of their lengths. This location of the greatest depths together with the below-sea-level altitude of the bottoms is indicative

Berby Hollow, seen from the slope of Gannett Hill (alt. 2,256 ft.) looking north. This is a characteristic, glacially eroded, eaves-trough type of groove across the upland of the western part of the Finger Lakes Region. Although wide and deep, the valleys shown serve only to collect and convey the minor local drainage of their sides and floors. In the climax stage of the glacierization they were major routes for the flow of basal ice currents. (Photograph by A. W. Abrams.)

Sixmile Creek through valley, near Wilseyville, New York, showing the glacially oversteepened valley sides and truncated spurs. At the summit the uniform level of the peneplain uplands is clearly in evidence. The flat floor of the valley is pictured near the site of the present-day divide between north- and south-flowing drainage. (Photograph by A. W. Abrams.)

Labrador Hollow through valley, showing the north entrance to the valley, looking south across the floor of the valley of the East Branch of the Tioughnioga River. The through valley here breaches the front of the Tully escarpment. In the left middle distance is the village of Apulia. (Photograph by O. D. von Engeln.)

Glacial Erosion

of the true rock-basin (closed-at-both-ends) nature of the depressions. Borings at the north end of Hemlock Lake struck rock 86 feet above the basin bottom; seismic soundings north of Canadice Lake indicate rock 39 feet above its bottom. Although there is no reason for presuming a possible down gradient southward for the valleys of the Finger Lakes "East," it is not impossible that the deepest rock excavation is south of the present heads of both Seneca and Cayuga lakes. A boring on the valley floor at Watkins in the Seneca valley went down 1,080 feet in loose materials; a boring in the middle of the valley a short distance south of the present end of Cayuga Lake struck rock at the depth of 430 feet. Allowing for a few feet of land elevation above the surface of the lake and fill on the bottom of the lake, this corresponds rather exactly with the greatest depth of the lake from soundings.

The available data for all the lakes are given in Table 1.

Table 1

	Lake	Altitude of surface above sea level, ft.	Greatest depth, ft.	Altitude of bottom below (−) or above (+) sea level, ft.
"East" lakes	Seneca	444	618	−174
	Cayuga	384	420	−36
	Owasco	710	177	+533
	Skaneateles	867	287	+580
	Otisco	784	66	+718
"West" lakes	Keuka	709	186	+523
	Canandaigua	989	274	+715
	Honeoye	818	?	?
	Canadice	1,099	91	+1,008
	Hemlock	905	96	+809
	Conesus	818	59	+759

In view of assigning the representative Finger Lakes "West" to glacial sculpture of the headwaters of the preglacial south-flowing drainage, originating in the uplands margining the Portage escarpment, it is appropriate that their most rugged sections should be at their northern ends. In assessing the manner and measure of this glacial sculpture a large number of parallel developments in other parts of the world are available for comparison and correlation of distinctive characteristics. The functioning of all mountain-valley glaciers essentially duplicates the descent of the continental glacier into the preglacial valleys of the Finger Lakes "West." The Swiss-Italian lakes, the lakes of the Lake District of England, the Argentine lakes on the east slope of the Andes are notable instances of lake basins excavated by glacier tongues descending through valleys from summit-level sources. Many dry valleys such as Yosemite and others in the Sierra Nevada were similarly fashioned by glaciers. The fiords of Alaska and British Columbia, of Chile and New Zealand still have at their heads the upper parts of the glaciers that made them. It could be objected that these are not strictly analogous occurrences in that these glaciers followed confined courses and did not have an ice flood of enormous thickness pouring broadside into and over them. But the fiords of Greenland now and of Norway during the glacial epoch were so conditioned and exhibit the same features as the lake-basin valleys of mountain glaciers. The Saguenay fiord opening into the St. Lawrence Valley is a parallel case.

Otherwise stated, whether the ice movement is upvalley or downvalley the erosional results are the same. The narrow sections of the former stream valleys are most drastically modified because massive directed ice currents in effect surge through them. Their floors are deepened, their sides steepened and straightened. Projecting spurs are cut away and truncated. Mountain valleys, formerly occupied by great glaciers but now clear of ice and free of glacial debris, on their floors show a cross section in rock that is exactly duplicated by that of the common type of eaves trough. Accordingly, it is probable that continuing the oversteepened slopes under the lake sur-

Glacial Erosion

face to reproduce this form would give the actual shape of the rock basin in which the waters are contained.

A notable characteristic of glaciated mountain valleys is a pronounced deepening of the main valley where a tributary valley, also formerly occupied by a glacier, joins the main valley. Seemingly the abrupt accession of the volume of ice brought by the tributary results in an immediate deepening of the main trough that may be likened to the hollowing out of a plunge pool at the foot of a vertical waterfall. This phenomenon appears in Keuka Lake at the point of junction of its two arms.

Despite the clear evidence of the effectiveness of glacial erosion in the excavation of rock basins in the floors of downsloping valleys, such action is apparently less potent than where the reverse is true. The basins of the Finger Lakes "West" are smaller, shorter, shallower than those of the Finger Lakes "East." In part this may be due to the circumstance that the uplands at the northern ends of the Finger Lakes "West" were appreciably higher than those bordering the northern ends of the Finger Lakes "East" so that the thickness of the continental glacier ice pressing into the downsloping valleys was less than that advancing into the north-sloping valleys of the Finger Lakes "East." Further, in the Finger Lakes "West" the ice flow was eased as the valleys opened out in their lower courses, whereas in the Finger Lakes "East" it was progressively constricted and forced through their narrowing heads.

Attention is thus again directed to the unmatched topographic situation of the region of the Finger Lakes "East" in relation to the advance of a continental glacier. As previously emphasized, no other region is known where a similar set of circumstances existed. If, accordingly, phenomena resulting from glaciation unlike those appearing elsewhere are encountered here, they may be regarded as developments owing to these exceptional conditions.

5 · Hanging Valleys and Through Valleys

THE case for attributing the origin of the basins of the Finger Lakes to glacial erosion and for asserting that they are rock basins with closed ends is set forth in the preceding chapter. By demonstrating that their excavation was through processes of ice erosion, a background is provided for interpreting two associated land forms that evidently are closely related—hanging valleys and through valleys.

One of the spectacular scenic features of a voyage through the Inside Passage to Alaska along the west coast of North America is the numerous foaming waterfalls descending the precipitous sides of the fiord walls. The streams furnishing these cascades emerge from valleys which have the eaves-trough profile that is the characteristic mark of a valley remodeled by glacial erosion. If one climbs to the lip of these side valleys, one finds they have flat, only gently sloping floors leading back to the summit regions of the mountain country paralleling the fiords.

The fiords, like the Finger Lakes basins, were glacially excavated and have been found through soundings to have similarly basined bottoms. They were the lines of trunk ice streams and owe their great depth and breadth to the superior erosive competence of the main ice currents. The tributary glaciers which formerly occupied the side valleys had only a fraction of the thickness of the trunk ice stream, hence were quite unable to keep pace in downcutting with it. Accordingly, when the ice was gone from both the tributary and main valleys, the tributary valleys were left "hanging" far above sea level in the main valleys and much farther above the rock bottoms of the main valleys. Farther north in the Alaskan fiord district there are identical features except that small glaciers still occupy the heads of the tributary valleys.

In these instances the relation is between *tributary* and main valleys. In the Finger Lakes Region there is a similar development of hanging valleys, but here it results from the relationship between main valley and *distributary* valley. In the southward mass movement of the continental glacier, the deep bottom ice developed internal currents that were guided by all the irregularities of the preglacial topography. The major volume of ice flow was concentrated in the valleys whose axes coincided with the major direction of glacier advance. These would be primarily the Finger Lakes valleys, especially those of Cayuga and Seneca. Other valleys would be utilized in accordance with the angle of their divergence from the north-south direction. For the most part such valleys would be tributaries of the main north-south valleys. A tributary coming in at right angles, that is, directly athwart the direction of the main advance of the ice, would facilitate only slightly, if at all, the flow of the glacier. It follows that, in proportion to the measure in which a particular tributary served to promote the southward flow of the interior ice currents, in that degree would it experience differential downcutting by glacial erosion. Further, once established as a channel way and the deepening started, both the volume of flow through the valley and the rate of its erosion would be progressively enhanced.

The early students of erosion in the Finger Lakes Region made much of the fact that the oversteepened slopes of both the Cayuga and Seneca valleys start uniformly at the altitude of 900 feet above sea level. This appears to have been the average level of the valley floors of the preglacial drainage and is indicative of the measure of their overdeepening. Before the concept of interior distributary currents and their erosive action was introduced, the early observers were troubled because upstream there was no regular progression of slightly higher hanging-valley lips. If the overdeepening by ice erosion was regarded as restricted to the main lake valleys, this would be the proper expectation. Instead the altitude of the lips was erratically distributed, some at the 900-foot level, others at different levels below this.

A fine view of various aspects of the hanging-valley phenomenon in the Finger Lakes Region is to be had on N.Y.S. Route 79 going from Ithaca to Mecklenburg at the summit of the curve of the road as it makes the ascent of the oversteepened Cayuga slope. Here one looks east into the Fall Creek valley flanked on the south by the northern front of the Portage escarpment. Fall Creek valley hangs at the orthodox 900-foot level. Its course is almost directly athwart the general direction of ice advance, and its valley was apparently not deepened by ice erosion.

Next, clockwise, is the Cascadilla Valley at a slightly higher level. This valley was cut off from the direction of ice advance by the front "range," Turkey Hill and Mount Pleasant, of the Portage escarpment. But the next valley, that of Sixmile Creek, trends south, hence was in line with the main ice flow and was a major channel for the general movement of the glacier. Its profile is seen to have the trough shape that is characteristic of ice erosion, and its floor is significantly lower, at 800 feet, than that of Fall Creek. Beyond the Sixmile Valley is the Cayuga Inlet. Since this was a direct continuation of the Cayuga Lake valley with its dominant ice current, the Inlet Valley experienced a comparable measure of excavation and does not hang. The floor of the Inlet Valley is on loose glacial, lacustrine, and alluvial deposits of unknown depth. The lake basin may well extend miles southward, for the oversteepening of the

Hanging and Through Valleys

valley sides here extends to the summit level of the uplands, 1,500 to 1,800 feet, instead of terminating at the 900-foot altitude.

A singular and impressive confirmation of this interpretation of the hanging valleys is afforded by Salmon Creek, tributary to Cayuga Lake on its east side, about eight miles north of Ithaca. The Salmon Creek valley is remarkable in that it is a relic of the long preglacial time when drainage in the Cayuga area was southward from the ancient peneplain divide, as was that of the Finger Lakes "West" region until immediately preglacial time. This is inferred because Salmon Creek is a hooked tributary to Cayuga Lake. Its course is south, whereas for a normal junction with a north-flowing main stream it should be northwestward. This is further confirmed by the forking of Salmon Creek into west and east branches, both directed southward appropriate to the preglacial drainage pattern. Because of the southerly direction of its course the Salmon Creek valley was a subordinate channel of ice flow, approximately in the same direction as the Cayuga Valley, hence was appreciably deepened by glacial erosion. In consequence, although its valley hangs, like those of other tributaries to Cayuga Lake, its floor was worn down to the low altitude of 500 feet. Its sides, below the 900-foot level, are over-steepened and are furrowed by a multiplicity of small short streams which start at the 900-foot altitude.

Tributaries to the Seneca Lake valley north of Millport hang at the same 900-foot altitude that evidently, as in the Cayuga section, marked the general level of the drainage basins of these two major north-flowing streams. Glen Creek, responsible for the carving of Watkins Glen, Rock Stream, and Big Stream are examples of hanging valleys on the west side; the streams at Havana Glen, Catlin Mill Creek, and Hector Falls are hanging on the east side. None of these is oriented to facilitate ice flow so they remain practically unmodified by glacial erosion.

The preglacial divide between north-flowing and south-flowing drainage was situated much farther north in the Seneca Valley than in the Cayuga Valley. The headwater erosion action of the Seneca drainage appears to have been less effective than that of the Cayuga drainage, perhaps because it was of lesser volume or because it encountered strata more resistant to stream erosion. In the Catherine Creek valley (inlet valley to Seneca Lake) the preglacial divide was evidently a mile or so north of Millport, because the valley is narrowest there and because from there southward the trend of tributary streams is southeastward and southwestward, as would be appropriate for drainage joining a south-flowing main stream. Johnson Hollow is an example. The floor of the Johnson Hollow stream hangs at the altitude of 1,000 feet, for the oversteepening of the sides and overdeepening of the main valley by ice erosion extend some ten to twelve miles beyond the preglacial divide site.

The hanging-valley relationship of preglacial tributaries exists everywhere in the Finger Lakes Region where glacially oversteepened slopes and overdeepened valleys are present. It applies to all valleys which were major channels of ice flow even though no rock basins were excavated. In some instances rock basins may be present but have been so completely filled with debris that the valley floors throughout are on loose deposits. If small lakes and ponds occur, they are shallow and result from inequalities in the surface of the deposits.

In its southward advance the general front of the continental glacier had eventually to surmount the Portage escarpment. In the province of the Finger Lakes "East," where the preglacial drainage was northward, the ice interposed a barrier to the escape of the north-flowing streams. It damned their valleys at their lower ends. Their waters in consequence were ponded and formed proglacial (in-front-of-the-glacier) lakes. The levels of these lakes, largely fed by glacier melt water, rose until they attained the height of the divide between the north-flowing and the south-flowing drainage. (Since the height of the ice of the glacier dam far exceeded the altitude of the divide, there could be no escape northward over the ice surface.) Then the lake waters found outlets across the lowest points in the divide and veritable floods poured down the steep slopes of the headwater streams of the south-flowing drainage.

Hanging and Through Valleys

The erosive effectiveness of this debacle could have been no less than profound. The deluge cascaded down the steep slopes of the narrow channels with the violence of a torrent. The rock formations were weak shales and thin-bedded sandstones cut by vertical joints. Both large and small fragments were readily dislodged by the force of the current. Even though the lakes and the overflow persisted only so long as it took for the glacier front to advance to the divide line, this period may be presumed to have lasted long enough for deep gorges to be cut across the divide heights.

This reconstruction is not altogether inference and deduction. Although in general all traces of such gorges have been erased by later ice erosion, there are a few sites where, at the *close* of glacierization, the ice front and topographic relations gave rise to ponding and overflow quite similar to those postulated above. One of these is Michigan Hollow, a stream-eroded rock gorge now occupied only by local drainge from its sides and head. (The gorge is bordered by an unnumbered road that leads southward from N.Y.S. Route 96 at the southern outskirts of the village of Danby.) The present stream, Michigan Creek, is clearly incapable of cutting a gorge of these dimensions. On the other hand, the flood of proglacial lake overflow water was not maintained long enough to finish the job of cutting the gorge, for near its head the descent is still at the rate of 64 feet per mile.

It is difficult to say just how the Michigan Creek overflow originated, but in the case of the rock gorge which is the outlet of Cayuta Lake the circumstances are quite clear. Projecting lobes of the retreating ice front dammed the low-altitude routes of escape of drainage from a wide tract surrounding the present basin of Cayuta Lake. Water from melting ice and precipitation collected behind the ice barriers until it attained a sufficient height to overflow at the lowest point in the rock rim of the basin. The gorge-cutting which ensued was so effective that a practically flat-bottomed channel at the level of the present Cayuta Lake was created. The dirt road through this outlet gorge has been abandoned, but it can be traversed on foot. (Leave Mecklenburg N.Y.S.

Route 79 over the unnumbered road leading south through Cayutaville to the outlet end of the lake.)

Both these gorges were eroded under conditions duplicating those postulated above as having brought about the transection of the preglacial divides between north-flowing and south-flowing drainage. But those transections were made at the time of the first ice advance, whereas the Michigan Creek and Cayuta outlet gorges were eroded as a last episode in the disappearance of the ice. Accordingly, whereas the divide-cutting gorges were almost immediately invaded by the advancing glacier, the relic gorges ceased to be even channels for voluminous stream flow as soon as the ice dam was dissolved.

If gorges similar to the relic gorges were developed in consequence of the initial advance of the ice, they were the precursors of the succeeding outstanding scenic phenomenon of glacial erosion—the creation of through valleys. The term, through valley, was supplied by W. M. Davis to designate wide, open level-floored passageways between opposed drainage systems. A pass is the equivalent type of gap across a mountain range. There is, however, one great difference between mountain passes and through valleys; in the former there is ordinarily on each side a considerable ascent to a definite summit level; in the latter the high point on the flat floor is commonly determined by the irregularities of the glacial deposits in the valley.

In the Finger Lakes Region the opposed major drainage systems connected by through valleys are the north-flowing Great Lakes drainage and the south-flowing Susquehanna River drainage. In keeping with their dominant status among the preglacial north-flowing streams it is appropriate and expectable that the largest, longest, and most representative of the through valleys are those prolonging the Cayuga and Seneca valleys into the Susquehanna drainage basin. The Cayuga and Seneca through valleys are far larger productions than would result from the mere reaming out of probable water-eroded transecting gorges. The great ice currents that eroded the rock basins of the two chief Finger Lakes swept

Hanging and Through Valleys

southward with such volume and such force as to carve out wide, deep, steep-sided troughs across broad belts of the former upland divide tracts. Along these cuts nearly vertical, oversteepened walls, counterparts of those margining the lake basins, extend all the way up to the 1,500-to-1,600-foot summit altitudes of the plateau south of the Portage escarpment. At their greatest height these nearly vertical walls rise 800 feet above the mile-wide valley floor of the Cayuga Inlet, and, if the loose fill under which the rock bottom is buried were removed, another 1,000 feet of elevation would, not improbably, be added to their precipitous slopes.

At the turn of the century when the idea of efficacious and large-scale glacial erosion was winning reluctant acceptance from geologists, the chief difficulty was unwillingness to recognize that the movement of the interior ice of glaciers is quite analogous to viscous flow. Glaciers were conceived, rather, to be rigid crystalline masses. This they are, down to a depth of approximately 200 feet below their upper surfaces. But an ice mass less than 200 feet thick does not flow and is, therefore, not a glacier. It was only when it became generally recognized that the physics of the interior ice was different from that of the exterior shell that appreciation of the tremendous erosive potency of glaciers was fully realized.

The straight, parallel walls of the Cayuga Inlet through valley betoken the competence of ice erosion to truncate the spurs that are characteristic features of the valley sides of winding stream courses. Such probably were preglacially present in the Finger Lakes Region. Even without an understanding of how this through valley was created, views along the Cayuga Inlet are spectacular. When seen as the product of glacier erosion by a concentrated current of the flowing ice, the wide trough and the precipitous walls are indeed imposing. They are best seen from the high point of N.Y.S. Route 34-96 between West Danby and North Spencer. The wide entrance to the trough as seen from the Cornell University campus on the road behind the Ezra Cornell statue is impressive when regarded in the light of its functioning in the creation of the through valley.

No such easily accessible, commanding viewpoints are available for the other great, fully developed through valley, that leading south from the Seneca Lake valley. In this instance the road, N.Y.S. Route 14, is on the floor of the trough, and the prospects are limited. The site of the preglacial divide, however, is rather definitely indicated by the constriction in width of the valley at Millport, and the truncated spur feature is prominently developed in the Horseheads region.

Although it was utilized by what may be regarded as only a branch of the Cayuga Lake ice current, the Sixmile Creek valley south of its hanging lip at Ithaca is a third large-scale expression of the through valley phenomenon. As seen from the Mecklenburg Road, N.Y.S. Route 79, the cross-section profile of the Sixmile Valley has the complete trough form characteristic of glacially eroded mountain valleys. From this it could be inferred that its floor is not deeply buried under loose materials, and that inference is confirmed by the exposure of rock at the level of its lip. The feature of oversteepened rock walls becomes prominent about five miles south of Brooktondale (N.Y.S. Routes 330 and 96B at and north of Wilseyville). That this section was the site of the preglacial divide is clearly indicated by the trends of the preglacial tributaries joining the trough from there on south.

Texas Hollow extends between Bennettsburg and Odessa (N.Y.S. Route 227 to Bennettsburg, 228 to Odessa). An unnumbered road on the floor of the valley connects these two villages. Texas Hollow is a full-scale development of the through-valley phenomenon. It is a mile wide, six miles long, has precipitous oversteepened sides 500 feet high and no definitely constricted section. Rock, however, crops out on its floor at about the middle of its length and two small streams start from there, one flowing north, the other south. Their sources at the rock outcrop may be presumed to be at the preglacial divide site. The dimensions and form, sides and bottom, of Texas Hollow are to be regarded as produced by ice erosion solely. Further, the sides swing in wide curves such as would

Hanging and Through Valleys

be quite appropriate to the flow through the valley of a current of stiffly viscous material.

In all aspects of form and development Texas Hollow conforms to the concept of the through valley. Its occurrence, however, is singular. There are no topographic elements immediately north of it that could be considered as fashioned to direct a major current of the interior ice of the glacier through this channel.

Although it is slightly outside, to the east, of the Finger Lakes Region, the Onondaga-Tioughnioga valley south of Syracuse (U.S. Route 11 from Homer to Nedrow) needs to be cited, since its dimensions exceed all the other through-valley occurrences. No lake appears in this great trough, because at the close of the last glacierization the inferred rock basin developed there was so completely filled with loose materials as to bring its surface up to the stream drainage level of the area. It is, in other words, probably a filled basin. The west side of the valley between Homer and Tully presents striking examples of truncated spurs. (Some years ago the staff of the Syracuse University Department of Geology promoted a project for seismic soundings along the axis of the Onondaga-Tioughnioga valley to determine the depth of the loose fill. Lack of funds precluded its execution. Such a series of soundings would be very revealing in regard to the longitudinal profile in rock of the through valleys.)

Comparable in occurrence with Texas Hollow is Labrador Hollow, south of Apulia, through which N.Y.S. Route 91 leads to Route 13 and Truxton. Like Texas Hollow, Labrador Hollow opens at the north into a broad east-west valley; unlike Texas Hollow it is in line with a large valley which no doubt fed a noteworthy volume of ice directly into it from the north.

A large number of other through valleys of small size (Halsey Valley, unnumbered road south from N.Y.S. Route 96 east of Spencer, for one) are all indicative of the flow characteristic of the ice, which permitted it to be guided and channeled into any topographic break that would facilitate its advance. A current once started in such a slot was immediately reinforced by the ice following and developed into a concentrated continuous flow. Even so, this grooving, wherever a portal was available, is impressive when a systematic field and map study is made of the phenomenon. Indeed, in places the dimensions of the passageways opened by minor lines of the flow are extraordinary. Thus at the head of Cascadilla Creek southward along Ellis Hollow South Road (comes into N.Y.S. Route 79 west of Slaterville Springs), there is a through valley over three miles long and a mile wide with a nearly flat floor that must have been eroded by a minor ice current moving up the Cascadilla Valley at right angles to the general direction of advance of the glacier and, in effect, finding an overflow route through the channel.

The final type of through valley to be considered is, on the other hand, an expectable and satisfying manifestation of the phenomenon. It occurs where, evidently, two considerable streams, one flowing north and one south, had, preglacially, directly opposed heads. It is even possible that competitive, preglacial headwater erosion had developed a col in the line of the east-west divide marking their sources. This topographic configuration was particularly suited for bringing about the gorge-cutting by overflow waters from a lake ponded in the valley of the north-flowing stream by the advancing ice front. Through these col-transecting gorges the rigid terminal ice was then thrust massively and in effect reamed out their walls. Then followed the current flow of the interior ice and its differential erosion.

There are two especially representative examples of this type of through valley in the province of the Finger Lakes "East." One is the Owego Creek valley extending from Dryden to Owego, N.Y.S. Route 38, which narrows progressively to just north of Richford, where, also, its walls are highest, and then opens out again southward. Rather obviously the site of the preglacial divide here was immediately north of Richford. The other example is the Tioughnioga River valley between Cortland and Marathon, U.S. Route 11, which, like the Owego Creek valley, narrows and widens and

Hanging and Through Valleys

has its most constricted section halfway between Blodgett Mills and Messengerville, where again the valley walls are highest and steepest.

Although rides through these two valleys will afford gratifying recognition of how completely their features fit into this interpretation of through-valley origin and history, the critical observer may have some reservations. For one thing the irregularly thick filling of loose debris, rock and soil, left behind when the ice finally wasted away, in places obscures the rock-profile form of the valleys. The other difficulty is revealed only through comparative study. The question is why the dimensions of these two through valleys, which were in line with the direction of ice advance and have broad feeding channels leading to them, are so small in comparison with troughs such as Texas Hollow and Ellis Hollow, which required a tortuous diversionary flow of the ice for their excavation. To this question there does not seem to be any ready answer, except perhaps that a number of minor parallel channels reduced the volume that the major troughs needed to accommodate.

The foregoing account of the through valleys may seem unduly long, but when it is recognized that these valleys perhaps are, even more than the rock-basin lakes, the distinctive hallmark of the region, this extended exposition will appear warranted. Although the cliffs of the through valleys lack the picturesque allure of the sunlit waters and verdant shores of the lakes of the region, they are, nevertheless, impressive scenic features in Central New York. Of course the through-valley cliffs are far inferior in grandeur to the canyon walls of the American West, but these are regularly seen against backgrounds of similarly magnificent proportions. Even a truncated spur may acquire a measure of glamour when viewed with a knowledge of the powerful forces operating in its creation.

Although the glacier penetrated to Williamsport, Pennsylvania, 119 miles south of Ithaca, and overtopped all the summits between these points, a circumstance that indicates an ice thickness of at least 2,000 feet in this area, I do not know of any hanging valleys, through valleys, or truncated spurs south of the Susquehanna

River. It would appear that the downward gradient of the Susquehanna drainage facilitated the ice flow sufficiently to preclude streaming through particular channels. If this is the correct interpretation, the Finger Lakes Region is further marked as a unique development of sculpture by the continental glaciers of the Pleistocene epoch.

Postglacial consequent gorge, eroded in the glacially oversteepened east side of the Cayuga Inlet valley south of Ithaca. It has bedrock floor and sides and is shallow, but carries a considerable volume of drainage. Compare with the photograph of the nearby interglacial consequent gorge. (Photograph by J. S. Wold.)

Interglacial consequent gorge. It has moraine-mantled sides and is deeper and wider than the postglacial gorge nearby on the same oversteepened slope. Despite its larger size, the volume of drainage is much less than that flowing down the postglacial gorge, although both pictures were taken on the same day. This inappropriate disparity of volume results from the postglacial diversion of considerable drainage, formerly directed to the interglacial gorge, to the postglacial successor. (Photograph by J. S. Wold.)

Rock pillar, Buttermilk Gorge, this erosion remnant was separated from the gorge wall to the left, of which it was once a part, by differential weathering and erosion. The rock zone where the path was made was less resistant to the disruptive processes than the rock in the pillar—though that looks very crumbly. This feature is representative of the variations in scenic aspect found in the different gorges. (Photograph from the Finger Lakes State Parks Commission.)

Fall Creek at Forest Home, New York, in flood. The major erosion of the postglacial rock gorges is done at such periods of high water. (Photograph by O. D. von Engeln.)

Ithaca Falls, Fall Creek. In postglacial time, 9,000 to 10,000 years, the face of the fall has moved back upstream only the distance shown in the picture, taken from a point at the end of the gorge. (Photograph by J. M. Anderson.)

Taughannock delta, 1952. The level area in the foreground is a delta deposit built up in Cayuga Lake of debris brought down by Taughannock Creek in postglacial time, about 10,000 years. (Photograph by Bennett Studio.)

Taughannock Gorge rock fall, 1935, from the side of the amphitheatre at the head of the gorge. The perfection of plane faces bounding the huge rectangular blocks indicates the smoothness and persistence of the joint and bedding planes in the structure of the rock walls of the gorge. (Photograph by O. D. von Engeln.)

Taughannock Falls in 1888, showing the fall with the projecting crest it then had. Sometime before 1892 the crest broke to the re-entrant angle now seen. (Photograph by E. M. Chamot.)

Taughannock Falls, showing the crest of the fall, with the re-entrant angle, as it has been since a rock fall between 1888 and 1892. Note the perfection of jointing in the Sherburne flags. (Photograph by O. D. von Engeln.)

Cavern Cascade, the only vertical fall in Watkins Glen. The rock behind the fall shows the smoothing produced by the grinding of sand-laden swirling water and is probably the upstream side of a great pothole, the lower side of which has been broken out into the downstream channel. (Photograph from the New York State Department of Commerce.)

Cascadilla Creek gorge. This marks the southern boundary of the Cornell University campus. Its whole picturesque length is easily accessible by footpaths. It is best seen going upstream. A geology class is in the middle distance. (Photograph by O. D. von Engeln.)

Chute channel in Enfield Glen (Robert H. Treman State Park). In flood the stream current lifts out and moves tabular blocks of rock downstream. The blocks are defined vertically by the perfectly developed joint planes. In consequence of this structural condition the artificial-appearing straight-walled channel has been eroded. (Photograph from the Finger Lakes State Parks Commission.)

6 · *Rock Gorges and Two Glacial Advances*

ALL the scenic phenomena of the Finger Lakes Region thus far considered—escarpments, cuestas, peneplain uplands, Finger Lakes "East" and Finger Lakes "West," rock-basin lakes, hanging valleys, and through valleys—depend in considerable measure on an understanding of their origin and natural history for their interest and appeal. For many persons they are not the eye-catching features of the landscapes.

Such is not the case with the rock gorges of the area. The mere mention of such features evokes in most people an emotional response, induced perhaps by mental images of romantic cliffs and foaming cascades. Aside from the lakes themselves, it is fair to say that the scenic glamour of the region derives in great measure from the almost innumerable chasms, some of truly spectacular dimensions, some so small as to be merely an embellishment of a village green.

The explanatory account that follows is based on the premise that continental glaciers originating in Labrador *twice* invaded New York State in such bulk and thickness as to overtop the highest peaks of the Adirondacks and Catskills and to extend along the Seneca Lake meridian as far south as Williamsport, Pennsylvania. Between the two advances there was an ice-free interglacial period, with a climate as warm as, or warmer than, that of the present, which lasted longer by far than postglacial time, which is now known to be of the order of 8,000 to 12,000 years. The whole of the Pleistocene glacial epoch apparently extended over one million years. From these figures it appears that the ice-occupation of the land during both advances must have endured for hundreds of thousands of years.

In the Middle West the evidence of multiple glacierization consists almost exclusively in the occurrence of superimposed layers of ice-deposited debris. These beds differ chiefly in the degree of decomposition through weathering processes that their materials have undergone. Where a profoundly weathered bed occurs under a fresh unaltered layer and both are composed of glacial materials, it is obvious that the lower layer is definitely older than the one overlying it. Further, between the times of the several depositions there must have been an interval sufficiently long to permit the decomposition of the lower bed to take place.

In the Middle West such superpositions are numerous. On the basis of their varying characteristics an elaborate stratigraphic sequence has been established, and four distinct and separate glacial advances are postulated. In view of these findings an intensive search for similar evidence of multiple glacierization has been made in the Finger Lakes Region. But aside from a few fragmentary deposits questionably interpreted as materials dating from the time of the disappearance of an ice sheet prior to the last one present, nothing resembling the Middle West superpositions has been discovered. It may be argued that the clear-cut successive beds of the Middle West are lacking here because the Middle West deposits were made on level lands at the outer borders of the ice sheets. In these belts the glacier was devoid of erosive force so that a new advance failed to clear away the debris left by its predecessor. Accordingly, it could be assumed that in the high topographic

Rock Gorges

relief of the Finger Lakes Region the strong current flow of the last ice scoured out all deposits made by earlier glacier advances and left only its own debris behind.

It is true that the last ice advance scoured the Finger Lakes country clean of all loose materials left behind by preceding glacierizations. Indeed, it is rather clear that the latest advance invaded a topography that had been so perfectly modeled by earlier glacial action to accommodate ice flow that no major relief adjustments needed to be made. The final glacier fitted exactly into the grooves made by previous ice work. Further, the magnitude of the last glacierization was less than any which came before it, so it may be regarded in a way as rattling around in its predecessor's boots. This is not to say that the last advance performed no rock excavation. It did, but the major sculpturing had already been done.

Despite these valid reasons for concluding that all depositional evidence of more than one glacial invasion of the Finger Lakes Region was cleared out by the last ice, there is clear proof of two advances, but of two only. This is afforded by the rock gorges of the area. Of these there are, as suggested above, many types, and it is rather intriguing that the smallest and least spectacular of such developments should have notable significance as evidence of only two glacierizations of the area. The more so because their import has not been recognized previously. These small gorges are conspicuous only when collectively noted, as where they are represented on a map. They are the channels of the short streams that descend the oversteepened sides of the lake and other glacially overdeepened valleys of the region. Technically these are consequent streams, namely, streams that flow down newly created slopes introduced in the landscape by processes other than ordinary weathering and stream erosion. How numerous and characteristic in occurrence these are is clearly shown on the map of their distribution around the lake basins.

Under field examination and study of relief maps the consequent gorges are noted to be of two types, distinguished primarily by being of different lengths. The distinctly shorter ones start immediately at the top of the oversteepened slope. On inspection these are found to be shallow furrows cut in the bedrock; miniature gorges whose sides and bottoms are completely free of glacial debris. The longer ones regularly have deeper and wider valleys and have cut back some distance into the upland above the oversteepened slopes.

Although the bottoms of the longer gorges may be in rock, their sides are mantled with glacial deposits. Where the topography was favorable to such action, the lower ends of these longer gorges may show an abrupt change to a steeper slope as though their former ends had been sliced off.

The shorter "clean" rock gorges are the ones that have come into being since the melting off of the last ice. The longer ones date from the time when the first glacier melted off the country. They were eroded during interglacial time. Because they are larger and longer than those cut postglacially, it is indicated that interglacial time was longer than the 9,000 to 10,000 years that have elapsed since the going of the last ice. That they were not completely erased by the erosional action of the last ice suggests that the first ice was responsible for by far the major part of the great glacial-erosion phenomena of the region.

The really big gorges—those that excite the interest and wonder of the sightseer—are the ones that have been eroded at the junctions of the larger preglacial east-west valleys with the overdeepened north-south lake valleys. These gorges mark the course of the large tributary streams in plunging from their hanging-valley lips down the oversteepened slopes to the levels of the present lakes.

Almost without exception these major gorges are postglacial, that is, have been eroded since the last ice melted off the land. In this respect they are like the short consequent gorges. Although the dimensions (depth and length) of the great gorges are really impressive, they, like the short consequent gorges, have sides and bottoms unencumbered by glacial debris. They are new cuts in the bedrock. Despite their similarity of origin and history the scenic aspects of the different gorges are sufficiently varied to hold the

Rock Gorges

interest of anyone who undertakes to explore a number of them in succession. When the sightseer has seen one he has not seen all. Differences in the composition and structure of the rock formations together with variations in the attack of the eroding streams provide novel prospects in each gorge visited. Their enduring attraction for the nature lover is ensured by this diversity of aspect.

The three outstanding examples of major gorges are Fall Creek, Taughannock, and Watkins Glen. They occur at the lake-valley ends of large east-west tributary streams that have retained most of the drainage area they had preglacially.

Fall Creek is the largest of the three streams. Its descent, 400 feet, from the level of the floor of the upper valley to that of the lake surface is accomplished in approximately one mile of its course. The gorge it has cut is eroded through the Ithaca shales and sandstones. None of the beds composing this rock section is sufficiently durable to function as a cap rock in the way the massive Lockport dolomite (Niagara limestone) does at Niagara Falls. Accordingly, the descent is not localized in a single vertical fall. Instead the stream goes down in a series of cascades and chutes. These all have a similar look. The aspect of the face of a cascade is determined chiefly by the spacing, depth of persistence, and trend of the joint-plane fractures which separate the bedrock into blocks. The falls front retreats upstream as such blocks are recurrently detached. The chutes apparently owe their initiation to the presence of persistent parallel joints that originally guided the current by facilitating block excavation in sequence along the line of flow. There is some evidence that the positions and altitudes of the successive cascades are related to former higher levels of the lake. A first cascade developed at the crest of the oversteepened slope as soon as this was exposed by the decline of the lake level to a point below its altitude. It then began a retreat upstream. At the sudden establishment of a next-lower lake level this process was repeated. Ultimately the lowest lake level, that of the present, came into existence, and the Ithaca Falls cascade, that seen from the Lake Street bridge, was started. This cascade has moved back only a short distance from its original position at the line of the oversteepened rock wall of the Cayuga Valley.

The Taughannock Gorge is quite different. The Tully limestone which crops out close to the present level and shore of Cayuga Lake provides a very enduring cap rock for the low vertical fall at the end of the gorge. The wide flat formed by the top of the limestone is distinctively marked by solution pits in its surface. These have been confused with potholes, holes worn in the bedrock by the circular swirl of sand and pebbles by a water current. But the irregular shapes and rough insides of the pits indicate their origin through differential solution of the limestone.

Upstream from the Tully fall the gorge floor has only a gentle gradient, but the sides of the chasm increase progressively in height until the site of Taughannock Falls, 215 feet high (55 feet higher than Niagara Falls), is reached, approximately two miles upstream from the Tully fall. The walls in the lower gorge are all black rock, the Geneseo shale, quite uniform in texture and of crumbly composition.

This shale is exceptionally yielding to mechanical attack. Accordingly, once the descent over the steep slope to each of the successive lake levels was started, a rapid retreat of the cascade face ensued. The weak Geneseo was worn away about as fast as it was exposed. Thus the lower course of the stream was eventually reduced to a low gradient fixed by the top of the Tully at the lower end of the gorge. The debris of this great excavation together with sediment brought from upstream was dumped in the lake at the mouth of the stream and now provides the broad picnic space of Taughannock Falls State Park. This expanse is a delta of quite the same origin as that at the mouth of the Nile, though of far smaller area.

The gorge excavation was halted where the outcrop of the Sherburne sandstones, at the base of the Ithaca formation, was encountered. Although these sandstones do not qualify as an ideal cap rock, they provided, by comparison with the peculiarly weak Geneseo, an effective barrier to a washout farther upstream. Thus the uniquely high Taughannock Falls came into existence.

Rock Gorges

The vertical mass descent of the water at flood periods has resulted in the excavation of a plunge pool 30 feet deep at the base of the fall, the largest development of its kind in the region. The pool is enclosed by a great rock amphitheatre, which forms the head of the main gorge. It was created by the action of spray from the fall in summer and the formation of ice in winter loosening and prying off the shale on both sides of the fall.

If the Sherburne sandstones were a better cap rock, they would be undercut by the enlargement of the amphitheatre so as to permit passage behind the fall at its base. But they do not hold up sufficiently well to permit this type of recess, commonly present in cap-rock falls, to develop. The relatively rapid retreat of the brink of the fall is indicated by the photograph, taken probably in 1888, which shows Taughannock Falls with a projecting crest line. Shortly after this picture was made, before 1892, the summit slab was dislodged and the fall crest became the re-entrant now seen.

Although Geneseo shale does crumble easily and rapidly where it is exposed to the weather, its interior masses are surprisingly coherent. This was demonstrated at the time of the disastrous flood of 1935, when a great rock fall occurred in the side of the amphitheatre to the left of the fall. The perfection of the jointing shown by the rectangular blocks composing the fall is remarkable, and from their appearance a rock of great durability might be inferred. But in a few years all this pile had crumbled, and no vestige of it remains.

The third notable gorge development, and the one best advertised and known, is Watkins Glen. Watkins Glen has been eroded in the Enfield formation. These rocks resemble the Geneseo shale of Taughannock Gorge in that they are rather homogeneous through a considerable vertical section, but they are far tougher and more durable than the Geneseo. Thin bedded sandstones alternate at short intervals with tenacious shales. Jointing is relatively inconspicuous.

In consequence of this structure, Watkins Glen, especially its lower end, is a narrow tortuous chasm through which the stream pours in twisted chutes. These slits are produced primarily by the scouring effect of fine sand carried by the current. The rocks margining the sides of the channels are smoothed as though they had been rubbed down with sandpaper. Above the levels where the grinding is currently effective, the rock walls roughen and spread progressively farther apart to the top of the gorge. This widening results from the piecemeal weathering of the rocks, greatest at the top because there they were first exposed to attack. At one point, above a pothole, the gradational transition from the mechanically polished rock to the roughened weathered rock is perfectly displayed.

Potholes are a noteworthy feature of Watkins Glen. They form where a rapidly flowing stream develops a circular swirl, presumably because of some initial irregularity of the rock bottom of its channel. The grinding action makes a hollow, and this catches the larger pebbles and boulders brought by the current in high-water periods. The biggest boulders fail to escape as the flood flow subsides and are thereafter rolled round and round, and so become tools for grinding a deep hole in the rock of the stream bed. Since the grinding is concentrated at the bottom of the depression it develops in the form of a broad-bellied pot. Several such potholes may be started close together; then in time they intersect each other at the bottom and the violence of a flood flow destroys the narrow rims separating them at the top.

There is one vertical waterfall, Cavern Cascade, of considerable height, about 50 feet, within the gorge. It is a cap-rock fall, and the crest has been maintained sufficiently long to bring about weathering of the rock behind the fall to permit passage behind the curtain of falling water (rather damp going). The Watkins Glen State Park authorities were dubious, however, in regard to the permanence of the brink, so they reinforced it with a mass of concrete.

At Rainbow Falls a small stream empties into the gorge from the top of its wall. The water is spread out so as to make a wide filmy curtain, and with sunlight in the right direction a rainbow appears in the misty screen.

Although smaller and less well known than the three described,

Rock Gorges

there are a number of other rather large gorges in the region. Each of them has its distinctive characteristics, and some of them deserve at least a passing mention here.

As Fall Creek gorge margins the northern side of the Cornell campus, so Cascadilla Creek gorge marks the southern border. Cascadilla merits description as a picturesque gorge. Its length is tree-lined, and it affords numerous cascading waterfalls, hence the name. It has been made accessible by paths at the waterside but has not lost its pristine beauty in a plethora of concrete construction.

Hiram Corson, early a professor of English in Cornell, an acquaintance of the poets Browning and possessor of a great flowing beard and a sonorous bass voice, came upon an undergraduate standing on the Central Avenue bridge over Cascadilla and gazing down the gorge. It was a bright day in October and the trees were flaming in their autumn glory of color. Corson stopped, caught the student's eye, then boomed, "The gorges are gorgeous!" The student gave him an instant's startled glance, then scuttled up the slope at a loping pace, presumably congratulating himself on a narrow escape from a character akin to the Ancient Mariner.

Two other gorges near Ithaca are, like Taughannock and Watkins Glen, state parks—Buttermilk Falls and Enfield Glen. (The lands for both of these were given to the state by Robert H. Treman, and the Enfield park is named for him.) Buttermilk is so called because of the long foaming cascade at its lower end. Its upper section is narrow and tortuous; a succession of potholes is a feature of the channel.

The upper reaches of both the Buttermilk and Enfield gorges are cut into the tough Enfield shale. But whereas joint planes are only inconspicuously present in Buttermilk, they have a major role in determining the distinctive characteristics of upper Enfield. There a perfect set of joints extends parallel to the stream course and is cut at short intervals by another set at right angles. This combination of fissures has permitted the quarrying out of huge rectangular blocks at times when the stream is in flood. In consequence the water pursues a course between straight, smooth rock walls in a chute so perfect as to give the impression that it is a manmade channel. Farther down the jointed structure has guided the opening out of a great amphitheatre, beyond which there is a long foaming waterfall like that at the lower end of Buttermilk.

These thumbnail descriptions should suffice to substantiate the assertion that each of the gorges in the region that have resulted from the overdeepening of the north-south valleys by glacial erosion has distinctive characteristics, hence is different enough to repay exploration. (See the section headed "Vantage Points and Excursions" at the end of the book.)

In earlier paragraphs it was also said that these gorges were all excavated postglacially. This is true, but the statement needs to be qualified. For instance, the terminal gorge of Sixmile Creek (Columbia Street bridge, Ithaca) is far wider than is warranted by the size of the stream; its floor is not on rock; its sides in places are mantled with loose debris; and it disappears at its upper end under a vast fill of deltaic and glacial deposit. At this point the wall of the terminal gorge is breached by the entrance of the stream from a postglacial rock gorge. A short distance up this side gorge there is a waterfall, and at its head the stream again flows over loose material and in a wide open valley. At low-water stages it can be seen that the stream is here pouring over the same wall of the larger gorge that the rock gorge breaches farther down. In other words, the postglacial side gorge here transects a rock promontory or outward bend in the course of the buried earlier, wider gorge. This is a situation several times repeated in the course of Sixmile Creek and often matched in most of the other gorge-forming streams.

Although they ordinarily do not have scenic expression other than a sag in the oversteepened slopes of the north-south valleys, debris-filled older gorges are known to parallel many of the postglacial gorges, and probably all are so matched. The filled gorges indisputably were eroded interglacially. They could not have been made until after the lake and through valleys had been overdeepened. The overdeepening was brought about by the first invasion of the continental glaciers. The cutting of the postglacial gorges is

Rock Gorges

still in progress. Accordingly, the older, generally debris-filled gorges must be assigned to interglacial time. Further, they are the only indubitable evidence of two glacial invasions of the region, and, since there is no clear evidence of a third set of gorges, only two glacial episodes are definitely authenticated.

Rather interestingly the older gorges are for the most part situated to the north of the postglacial cuts. It would appear that, when the second ice advance occurred, it shoved the interglacial gorges full of debris, then overrode this fill and continued over the unfurrowed rock surface on the south. Postglacially these rock surfaces proved to be the lowest lines available to the restored stream flow, hence the sites of the present gorges. Or, since the retreating ice front first bared the area south of the filled gorges to renewed stream flow, the postglacial drainage was established and persisted there.

Unburied sections of the interglacial gorges (as in Sixmile Creek) are regularly larger than their postglacial counterparts. From this difference it can be inferred that interglacial time was longer than postglacial time. Further evidence of such time difference is had from the circumstance that the interglacial gorges had extended their courses much farther upstream by headward erosion than have the postglacial gorges. In consequence of this the postglacial drainage tends to weave in and out of the lines of the filled interglacial gorge cuts. Such wandering gives rise to a succession of wide open amphitheatres, where the loose fill has been washed out, and secondary upstream postglacial gorges, eroded where the new stream course encountered rock at altitudes above the level of the top of the walls of the interglacial gorge.

This interweaving of postglacial and interglacial gorges, as in Sixmile Creek, not uncommonly involves even the lower courses of the streams. Thus Fall Creek at Triphammer Falls intersects the interglacial gorge course, which is represented by Beebe Lake. At the head of Beebe Lake a short secondary postglacial gorge leads back to the filled gorge whose near wall is just above the lower bridge of Forest Home. On the far side of the old gorge rock appears in the floor of the creek, and upstream from there no trace of the old gorge appears. The situation parallels that of the Sixmile occurrence previously described.

The same relationship appears in upper Enfield gorge on a grander scale. The amphitheatres above and below the intersecting gorge here are of greater size, the intersecting gorge itself is longer and more impressive. Downstream from the great cascade that terminates the major intersecting gorge, the courses of the new and old gorges are completely intertwined except for a final intersecting gorge with a waterfall and pool at the main-valley level. In Watkins the interglacial gorge is not in evidence except at the very head of the glen; at Taughannock it is nowhere exposed as such.

The recital of these upstream gorge developments could be indefinitely prolonged. Their listing would be merely boring. But as objectives of personal excursions they have great allure. The big, immediately accessible gorges are so much visited and exploited that they have lost much of their original wilderness charm. If, however, one cons the topographic sheets of the region, there will be discovereed many such rocky glens, remote and unspoiled, which can be enjoyed in their natural beauty, away from the crowd (see "Vantage Points and Excursions" at the end of the book). It is quite safe for me to make this suggestion, for only those imbued with the true exploratory spirit will make the effort required to search them out and have a share in the reward they offer.

There remain, however, two unique occurrences which need to be specifically described to round out this account of gorge occurrences. One is the intriguing gorge sequence associated with the outlet to Cayuta Lake; the other is the course of the Chemung River through what was formerly known as Roricks Glen, a term that survives only as the name of a Boy Scout Camp within its confines.

Back in 1893, when the first topographic survey of the Cayuta Lake area was made, the men who did the work with transit and rod made an assumption to save themselves work. It was a reasonable enough inference in a general sense, but the surveyors were

Rock Gorges

not sufficiently versed in geomorphic science to read the field evidence correctly. On N.Y.S. Route 13, about one mile east of its intersection with N.Y.S. Route 224, there is a long, low concrete culvert. This bridges a wide flat-bottomed channel leading up to a narrow gap in the steep and high rock slope that is here the side of the large valley followed by Route 13. The 1893 surveyors assumed that this gap was the lower course of a hillside gulch that had its head at the crest of a ridge and that it was matched by a similar valley down the other side of the ridge; and that was how they mapped it. What the surveyors failed to recognize or appreciate was that the wide flat-bottomed channel leading out from the rock gap was a wholly inappropriate discharge form for a gully stream coursing down the hillside. Actually the rock gap is the lower end of a transecting gorge which at the close of the ice occupation of the region carried the overflow waters of a glacially ponded lake of large dimensions occupying a basin of which the present Cayuta Lake is the center. The gorge outlet of Cayuta Lake now is its exact counterpart or, better, is the upstream member of a triad of such gorges. Because it escaped the mapping activities of the early surveyors, the downstream unit has been dubbed the "Lost Gorge."

At a time when the broad valley followed by Route 13 had been cleared of ice at the close of the last glacierization, the glacier front rested inertly for a considerable period in a land bay to the west and south of the Cayuta basin. This ice made a completely effective barrier to the escape of drainage from the Cayuta area by what was, under the ice, the lowest downgrade overland route. Behind it the melt water and ordinary precipitation were ponded. The water could not escape northward because the unbroken glacier was still in occupation there. Accordingly, the level of the ponded waters rose until it reached the level of the lowest gap in the east-west ridges east of the ice-occupied land bay, and there it overflowed.

It is difficult to determine the sequence of the erosion of the triad of gorges by the overflow waters. Very probably the "Lost Gorge" farthest south was cut first. The Cayuta shale, which is the

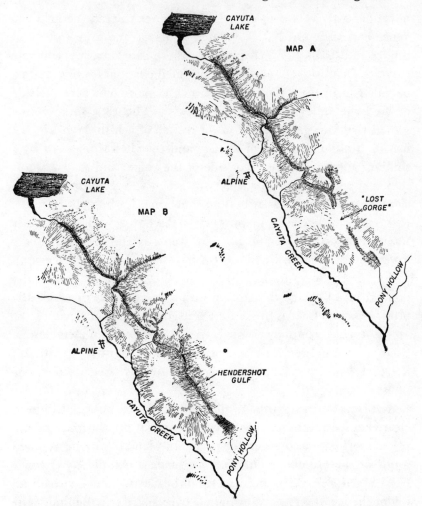

Map A. Cayuta Lake outlet gorges, as mapped by the United States Geological Survey topographers in 1893. The presence of the lowest gorge, now named Hendershot Gulf, was not detected, hence the sobriquet "Lost Gorge."

Map B. Cayuta Lake outlet gorges, as actually developed and as mapped in 1944–1950. This shows the existing drainage, and the "Lost Gorge" appears, named Hendershot Gulf. (Drawings by R. Andrews.)

bedrock of the area, is extremely weak and would succumb rapidly to the rush of a large volume of water down a steep slope. The level of the original col would be shortly reduced below that of the col in the ridge next north, and then the cutting of the second gorge

Rock Gorges

Glacial Cayuta Lake and outlet gorges, first stage, showing a reconstruction of the conditions that obtained when Cayuta Lake was first ponded by the barrier of glacier ice blocking the preglacial drainage of its basin toward the southwest. Initial erosion of the "Lost Gorge" and of the existing outlet gorge are shown. (Drawing by R. Andrews.)

would begin; and so, in turn, with the third gorge, which still serves as the outlet of the lake. All three gorges are quite flat-bottomed—the perfection of the transections achieved is remarkable. The conjunction of circumstances that brought about this multiple gorge development is shown in the accompanying figures. The original mapping of the gorges is paired with the aerial photo-mapping of 1950 on which appears the name Hendershot Gulf for the "Lost Gorge." Despite this new identification by the local place name, this gorge still has an aura of the unknown and mysterious. For those with a zest for exploration it is suggested that they undertake the passage in early winter after the ground has frozen, when firm ice sheets extend over the swampy spots. Even so, waterproof footgear is recommended. It is somehow comforting to conclude that the "Lost Gorge," under any name, will never become a concrete-equipped public park. It would be gratifying to have it made a natural history preserve, so that the forest on its slopes and floor would always be spared the lumberman's attack.

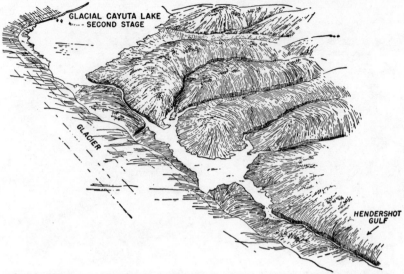

Glacial Cayuta Lake and outlet gorges, second stage. The "Lost Gorge" (Hendershot Gulf) has been deeply enough eroded to drain both the original lake and the intermediate lake to a lower level. At a later stage, yet while the ice barrier still persisted, the middle gorge, here under the waters of the intermediate lake, was eroded. (Drawing by R. Andrews.)

The other unique gorge development referred to above, the course of the Chemung River through what was formerly known as Roricks Glen, on N.Y.S. Route 17E between Big Flats and Elmira, is a production of the first ice advance, preserved through interglacial times and the second glacierization and, in effect, reinstated in its functioning when the second glacier melted away.

Preglacially the Chemung River flowed through the wide open valley between Big Flats and Horseheads. At Horseheads it was joined by a small tributary from the north, occupying what is now the southern end of the through valley from Seneca Lake. In the first advance of the glacier the ice current which created and poured out of this channel spread laterally when it got to Horseheads and dammed the eastward flow of the Chemung. The waters so ponded rose in level until they found an overflow into the broad valley between Horseheads and Elmira through the lowest col in the hill between the upper Chemung course eastward and its southward continuation. There was again the rush of waters

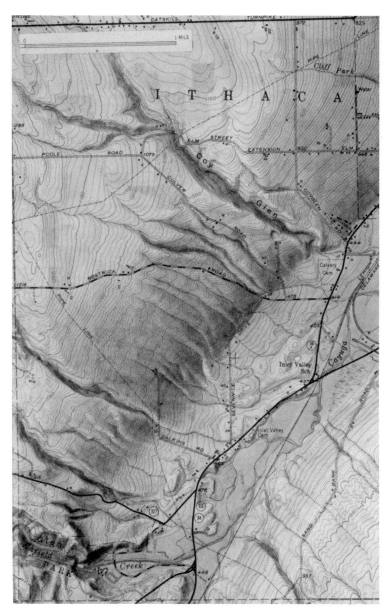

Coy Glen terraces ("The Steps") and interglacial and postglacial gorges on the glacially oversteepened slopes of the Inlet Valley. This halftone reproduction of part of the Ithaca West, shaded-relief, topographic sheet shows the high-level proglacial-lake delta terraces of Coy Glen, the highest at 1,040 feet (just below "Elm Street" on the map), which is the outlet level of Lake West Danby. Also shown, southwest of Coy Glen, are a number of consequent streams, of which the shorter ones are inferred to be postglacial, the longer ones interglacial. (From the Ithaca West quadrangle map, U.S. Geological Survey.)

Cayuta Lake outlet gorge, showing the floor of the middle gorge of the series of three, now almost completely devoid of drainage. The picture suggests the wilderness aspects of these gorges. (Photograph by O. D. von Engeln.)

Ice cliff of a glacier terminating in deep water. This picture of the front of an Alaskan glacier ending in the sea well represents the probable appearance of the ice barriers that dammed the waters of the high-level proglacial lakes of the Finger Lakes Region. (Photograph by O. D. von Engeln.)

Rock Gorges

down the steep slope beyond the col and the rapid transection of the hill by a water-eroded gorge. Since the Chemung drainage is liable to floods, this cut was probably made apace. Then the advancing ice overtopped it and the bordering hills and swept southward.

When the first glacier had melted back to the site of the former ponding, a low-level escape for the Chemung waters was available through the new gorge, even though this had been partly filled by deposits from the wasting ice. The Chemung remained in this course even after the ice plug at Horseheads had melted away because thick gravel and sand deposits made by streams from the ice front raised the land level at Horseheads above that at the head of the gorge. Then, all through interglacial time, the Chemung, mainly during periods of flood, greatly enlarged the gorge, widened its floor, and modified its course to the curved pattern appropriate to normal stream-valley development.

With the return of the ice in the second glacier advance, the sequence was different in that the whole Chemung drainage was simply displaced and the gorge overridden. But, interestingly, the ice fill of the gorge valley seems to have protected its south wall from glacier erosion and thus preserved its ragged rock top, interglacially weathered. It is possible, however, that this weathering is postglacial. Thus, when, after a second wasting of the ice, the Chemung was restored to its interglacial course, it had only to contend with some new glacial debris covering the floor of the valley.

If deposits of moraine made by the first glacier are preserved in the Finger Lakes Region, it may be they could be found in this Chemung gorge valley under those left behind by the second ice. To uncover them, however, would require laborious and expensive excavation, and it would be difficult to fix on a site where the chance for success is best. But even without such confirmation of the sequence of events here outlined, it will presumably be appreciated that the Chemung development is a unique feature of the Finger Lakes landscape, well worth a visit by perceptive students of natural history.

7 · The High-Level Finger Lakes

WHEN an observer with eyes sensitive to land-form differences looks across the Cayuga Inlet valley to its west side from a point on N.Y.S. Route 13 about one mile south of Ithaca, he becomes conscious of a notably distinctive feature of the landscape.

Opposite this point on the road the mouth of the gorge of Coy Glen is marked by a series of terraces separated by steep slopes. These terraces were sufficiently conspicuous to have attracted popular attention early and became known as "The Steps."

In all, seven terraces may be discerned. Each terrace represents a former higher level of a lake in the Cayuga basin. The material composing the steps is gravel and sand, sediments brought by the Coy Glen stream to the high-level lakes and deposited at its mouth. In this manner a delta was built at each level. Each delta, in turn, was transected by the stream as the lake levels were successively lowered. Thus there are now parallel flights of stairs, one on each side of the gorge. By analogy with the term hanging valley these are fittingly referred to as hanging deltas.

The High-Level Lakes

Although the Coy Glen hanging-delta exhibit is as near perfect a showpiece of the phenomenon as could be wished, the accumulations there were those of a small short stream. Delta deposits of far greater volume were made where a large tributary stream entered the high-level lakes. Of these, two—Fall Creek at Ithaca and Glen Creek at Watkins Glen—have first importance.

It may be presumed that most literate inhabitants of the Cayuga and Seneca lakes areas have heard references to the earlier higher-lake levels. It also appears that interest in the reasons for their existence is second only to that manifested in the question of the origin of the gorges.

The basic reason for their development is rather simple. Without exception the overflow of the Finger Lakes is now northward and is a part of the Great Lakes–St. Lawrence River drainage. Any barrier that could prevent this northward escape would bring about a rise in the level of the lake waters and, when the waters were ponded high enough, would bring about overflow either east, south, or west by outlets at higher altitudes than those which now function. The front of the continental glacier was such a barrier during all the time that it extended east-west across the country on a line north of the present divide between the Great Lakes–St. Lawrence River drainage and the Susquehanna River drainage. Since the ice then existed as an uninterrupted mass from its southern terminus to the interior of Labrador and rose to altitudes approaching 10,000 feet, there was no outlet level lower than those of the divides leading down into the Susquehanna River drainage. Accordingly, as soon as the ice front melted back in any valley sloping northward from the Susquehanna divide, water was dammed to form a lake between the ice barrier and the divide level. Once it reached the divide level, it overflowed and became part of the Susquehanna River drainage.

Thus far the story is easily understood. Complexities are encountered when it is sought to discover just how high each of the eleven lakes thus formed would need to rise in order to overflow its particular divide height. After that, the problem is to learn how the lakes, at first separate at the southern end of each valley, got linked up across country (east-west) to form combined lakes at successively lower levels and, in general, of greater extent.

The tabulation which follows lists the Finger Lakes in order from east to west with the present water level of each and the altitude to which this would need to rise in order to have a first southward overflow to the Susquehanna River drainage as indicated by topographic maps.

Name of lake	Present level	Overflow level
Otisco	788	1,300
Skaneateles	867	1,300
Owasco	711	1,060 into Fall Creek
Cayuga	384	1,040
Seneca	444	900
Keuka	709	1,140
Canandaigua	686	1,400
Honeoye	803	1,189
Canadice	1,076	1,200 into Hemlock Lake
Hemlock	905	1,040
Conesus	818	900

It should be realized that such a sequence of lake history accompanied both ice advances and each melting off of the glaciers. The transecting streams, described at some length on preceding pages, were born when ponding due to the first ice advance took place. Except for traces of the gorges then eroded, all evidence of the existence of high-level lakes previous to those associated with the melting back of the second ice is gone. The shore lines and deltas of the lakes of the first advance, the first retreat, and the second advance were successively erased by ice erosion. No doubt even the levels of the spillways were changed. So the record now

The High-Level Lakes

available is solely that of the lakes held up by the glacier dam when the second ice was melting away.

Since Lakes Cayuga and Seneca extend farthest south now, their basins obviously harbored the first of the high-level lakes to develop as the ice front was shifted northward for the second time. Actually these first lakes were initiated in the through valleys which are prolongations of the Cayuga and Seneca basins. Further, the divide sites were in-valley, that is, valley bottom elevations composed of glacial deposits. At such sites the major divides in the Cayuga and Seneca basins, and those elsewhere in the region, have since remained.

The initial high-level lakes were small ponds. Their overflow volume, however, was relatively great because they were fed by melt water from the ice lobe projecting up the valley. At the stage when the lakes came into being, the summit tracts along the east-west line marked by the termini of the valley lobes were already clear of ice. Over the hills the thickness of ice was necessarily less than in the lowlands; hence a given amount of melting would remove the ice entirely from the summit areas while massive ice tongues still occupied the valleys. Besides, as has been emphasized on preceding pages, the north-south valleys were the main channels of ice flow, hence were the lines along which the glacier front was supplied with the greatest volume of new ice.

Despite the probable large volume of overflow from the lakes when they were first ponded behind the in-valley divides, there was no rapid erosional deepening of the spillways. The gradients to the south over the valley floors were mostly gentle. Further, lake overflow waters are practically devoid of coarse sediments; gravel and sand sink to the bottom of the lake when the current that supplies them encounters still water. Accordingly, a typical lake overflow channel in loose materials is broad and flat-bottomed with low but steep banks. Such canal-form channels are made by the sweep of the current, not by the grinding erosion of sand and rock tools. The Niagara River upstream from the rapids above Niagara Falls is the climactic example of such a development. At the other end of the scale the phenomenon is nicely illustrated by the channel at the outlet of the "Lost Gorge" (now Hendershot Gulf) of the Cayuta Lake overflows. An intermediate instance is that parallel to U.S. Route 11, the spillway of the lake ponded by the ice front in the Onondaga Valley, seen a short distance north of Tully, where it is dry, and south of Tully as the West Branch of the Tioughnioga River. In the case of the Skaneateles high-level-lake overflow, the start (N.Y.S. Route 41) is obscure but conforms to the pattern. The descent there, however, is steep so that, after a few swings back and forth across the road in a channel of the typical form, the original overflow current became a torrent and eroded a regular valley trench.

These examples should suffice to establish the concept and characteristics of lake-overflow channels in unconsolidated materials. With them in mind, readers imbued with the explorer's urge may find it rewarding to search out the highest spillways of the other Finger Lakes. Regrettably, in a sense, those serving the initial lakes in the Cayuga basin are less well defined, and less readily discernible than those cited above.

Probably the first of the high-level lakes to form was the one at what is now the source area of the Cayuga Inlet, the in-valley divide site of the Cayuga-Spencer through valley (about one-half mile north of North Spencer on N.Y.S. Route 34–96). The actual overflow was at the elevation of 1,040 feet, eastward through a rock gap into the southward continuation of Michigan Hollow. The rock gap has the flat-bottom characteristics of overflow channels, but this cut may have been started by a stream flowing along the margin of the ice lobe just before the ice front had melted back far enough for the first lake to form. This lake, called Lake West Danby, was confined to the upper Cayuga Inlet valley and persisted until the ice front had been wasted back as far as the nose of South Hill at Ithaca.

Although Lake West Danby was the first of the high-level lakes to appear in the Cayuga basin, there developed later, probably next in the succession, a lake with a higher altitude of overflow, 1,270

The High-Level Lakes

feet, which occupied the valley of the eastern branch of Sixmile Creek, called Slaterville Lake after the village (Slaterville Springs) whose site it covered. Slaterville Lake emptied into the south-flowing West Branch of Owego Creek and perhaps extended through Ellis Hollow (divide level 1,060 feet) into the upper Cascadilla Creek valley. Slaterville Lake was probably shallow because the valley bottom of Sixmile Creek up to the point where the stream turns northeast (east of Slaterville Springs) and the tract between this elbow and the similar elbow of the West Branch of Owego Creek are floored with glacial deposits of various types up to 1,250 feet in altitude.

While Slaterville Lake existed at its highest level, 1,270 feet (the West Branch Owego Creek overflow), the ice barrier pressed against the steep north slope of Bald Mountain south of Brooktondale and extended up the southern, Wilseyville, arm of the Sixmile Valley. When the ice had melted away from the face of Bald Mountain below the 1,270 foot elevation and the level of the ice choking the southern extension of the Sixmile Valley was also reduced below that level, Slaterville Lake was first drained down to 1,070 feet by a channel in the shelf at the base of Bald Mountain, which starts at that altitude. The escape of this water was southward toward Wilseyville, evidently over the surface of a stagnant ice block filling the bottom of the valley. This surface appears to have progressively declined in level until it got down to approximately the 1,040 foot altitude, at which elevation the western outlet of the Sixmile Creek valley was also cleared of ice. Then the Bald Hill spillway was abandoned and Lake Slaterville was drained completely.

At this time massive delta deposits were made at Brooktondale with their tops at 1,040 and 1,020 feet. Unlike the sediment-free waters of a lake overflow, the waters which made the Brooktondale deltas were loaded with coarse materials brought by Sixmile Creek. Since this stream was then still being fed by melt water from the glacier front bordering the valley on the north, it had a volume far larger than the present average drainage. Besides, its course was over the soft gravel, sands, and mud of the newly exposed lake-bottom. Therefore, large quantities of detritus were swiftly transported and dumped at the mouth of the stream into the new lake, first at the 1,040-foot level, then at the 1,020-foot level.

It is now recognized that singly and collectively the ice-dammed, high-level Finger Lakes had, geologically speaking, a very ephemeral existence. Although the deltas built into them are referred to above as massive, their volume actually is very slight in comparison with what could be expected from long periods of accumulation. The whole series of such lakes endured only so long as the time required to melt the ice front back some fifty miles or so. Estimates of the rate of the retreat of the ice front, supported by evidence that is at least indicative of the order of magnitude of the phenomenon, arrive at 200 feet annually as the approximate figure. This would provide about 1,000 years for the whole history of the proglacial Finger Lakes. For comparison, the time since the ice was present in the region is computed to be at least 9,000 years. In that time the shores of the Finger Lakes at the present levels have acquired high rock cliffs by wave erosion and enormous delta deposits at their heads and also deltas from side-tributary streams.

In consequence of the, by contrast, fleeting existence of the high-level lakes, the shore-line features marking their altitudes are faint, obscure, and very discontinuous. Such failure to leave a mark of their presence should be especially true of the highest level attained in any of the valleys, because these were the smallest lakes of the succession and probably shortest lived. But their altitudes are more readily determined than those of later lower stages because of the abrupt transition at their shore lines from sediments (sand and clay deposited on the lake bottom) to the unassorted glacial debris laid down directly by the ice. Such a transition is also in evidence at each of the succeeding lower stages northward from each line where retreat of the ice front resulted in the uncovering of a new outlet at a lesser altitude than that previously operative.

The ephemeral existence of the earlier stages of the lakes lends

The High-Level Lakes

support to a concept of stagnant ice covering the bottoms and plugging the mouths of valleys like that of Sixmile Creek. As such inert ice masses were rapidly and progressively dissipated, lake levels could be lowered both gradually and abruptly until stability was established by overflow across the lowest land divide between the north-directed drainage and the Susquehanna River drainage.

In the Sixmile Creek valley (Wilseyville branch) this land spillway was at 980 feet and is located in a swamp north of the settlement of Whitechurch. The waters first ponded at the 980-foot level are called Lake Brookton.

Meanwhile the front of the ice lobe occupying the Cayuga Valley was still pressing against the nose of South Hill (Ithaca). Thus it prevented the waters of Lake West Danby, emptying at 1,040 feet at North Spencer, from flowing around the prow of the hill into the Sixmile Valley lakes. But sometime after the 1,020-foot level had been established for these lakes, a channel between the glacier front and the rock of the hill was freed of ice at an altitude less than 1,040 feet, and Lake West Danby was drained down first to the 1,020-foot Slaterville Lake level and then down to the 980-foot level of Lake Brookton. A slight farther melting back of the ice front permitted the merging of Lake West Danby and Lake Brookton at the 980-foot level, and Lake Ithaca—the joined lakes—came into existence.

There is a cult among geologists interested in the high levels of the proglacial lakes which adheres to the concept that a postglacial northward uplift of the land made the original horizontal shore lines of the lakes rise toward the north. This conclusion in the Finger Lakes Region is based primarily on correlations between levels of delta tops at successive occurrences south to north. Along the shores of the Great Lakes, especially Lake Michigan, where long continuous high-level beaches exist, such tilting is conclusively demonstrated. But, as noted above, the high-level Finger Lakes shore lines are ill defined and discontinuous. Because of this obscurity, the matching of the delta-top elevations is quite unsure. Besides, if the hinge line of the Great Lakes tilting is north of Central New York, the Finger Lakes levels would not be affected.

Hence major dependence in fixing levels is by the transition line between sediments deposited in a lake and unassorted glacial deposits. After that, the altitudes of outlet channels and the heads of overflow channels are eminently definitive. Thus the overflow channel around the nose of South Hill starts at 1,040 feet, and this elevation corresponds exactly with that of the outlet level of Lake West Danby. There seems to have been no tilting throughout the length of the Cayuga Inlet valley.

Again, the outlet level, 980 feet, of Lake Brookton-Ithaca matches exactly the transition line between lake sediments and glacial deposits in the Cascadilla Creek valley, as seen on the Ellis Hollow Road. At this locality there is a rather well-defined shore line, and even a low wave-cut cliff was developed in the rock of the hillside. The 980-foot Lake Ithaca level is further confirmed by a shore line at precisely that level on West Hill (Ithaca) where it is crossed by N.Y.S. Route 79. At the 980-foot level all the campus of Cornell University and the lands beyond as far east as the base of Turkey Hill were under water.

(Since 1940, the United States Geological Survey has made available new topographic maps of the whole Finger Lakes Region on a scale of approximately 2.75 inches to the mile and with 10- and 20-foot contour intervals. The larger scale and, where used, the 10-foot contour interval permit representation of details that the earlier, one-inch-to-the-mile, 20-foot contour maps, surveyed as long ago as 1890, could not show. Since the new maps are based on aerial photographs, they are completely objective. Presumably the altitudes on the new maps are also more accurate than those of the earlier ones. The reader who wishes to check on the field evidence here adduced should have copies of the new large-scale maps in hand.)

The concept of control of lake levels by stagnant ice blocks is borne out by the glacial-lake history of the lower Fall Creek valley.

The High-Level Lakes

That a well-nourished massive ice lobe of the Cayuga Valley ice stream was protruded up the Fall Creek valley in the closing stage of the glacier occupation of the region is established by the enormous deposit of ice-transported debris outlining its terminus at the village of Varna. The ice-contact side of the deposit is the so-called Varna Hill of N.Y.S. Route 13. From the hill the deposit swings in a great crescent northward and eastward across the Fall Creek valley. Until this lobe became inert and melted down, there could be no lake in the valley area between Varna and the main ice stream in the Cayuga Valley.

As soon, however, as the ice mass in the Fall Creek valley had melted down below the 980-foot level, the Lake Ithaca waters entered the area from the south. At this stage a puzzling collection of data is encountered. The extension of Lake Ithaca across the Fall Creek valley is confirmed by the utilization of an overflow across its north bank at the critical altitude of 980–970 feet. This spillway evidently succeeded the 980-foot overflow in the Sixmile-Wilseyville valley and functioned for an appreciable period. The exact overflow site is now the pond and swamp that is the source of Pleasant Grove brook. (It may be visited, going east, by turning off N.Y.S. Route 392 at Forest Home into Warren Road; then follow the first road to the right off Warren Road one-half mile to the site.) In its upper reach the channel leading away from the spillway, now followed by Pleasant Grove brook, has the characteristic lake-outlet, cross-section form, wide flat bottom, shallow (2 to 5 feet deep) with steep banks. It leads down to wide delta tops at the 940-foot altitude, now in part occupied by Pleasant Grove Cemetery.

After the 940-foot level of Lake Ithaca was established and after the Pleasant Grove overflow was functioning, an escape along the line of the Cascadilla Creek valley became available and drained all of the 980-foot-level Lake Brookton waters down to the 940-foot Lake Ithaca level. At this time the 940-to-930-foot delta of Cascadilla Creek at East Lawn Cemetery of East Ithaca was built. Concurrently, the Pleasant Grove overflow was abandoned.

There remains the problem of how the second, lower, 940-foot level of Lake Ithaca was determined. Since all the avenues of escape southward have been canvassed and assigned appropriate roles in the sequence of levels, and since lands to the east are significantly higher than the 940 feet, the only remaining possibility is a westward outlet. If this was the route, then contact must have been made with the high-level waters in the Seneca Lake basin.

In comparison to that of the Cayuga basin the proglacial lake history of the Seneca basin is simple. The difference is due to the low altitude, 900 feet, of the initial escape level of the waters ponded in the Seneca Valley. The overflow here was at Horseheads into the Chemung River, tributary to the Susquehanna. Because of this low-level overflow, the ice-dammed lakes in the Seneca basin did not invade any tributary valleys. In fact they barely reached to the top of the oversteepened, ice-eroded slope of the main valley in the part now occupied by Seneca Lake. The initial lake in the Seneca Valley has been named Lake Watkins.

Lake Watkins at the 900-foot level was the only escape route available to the Cayuga basin drainage below the 980-foot level as long as the ice barrier persisted in the middle reaches of the modern Cayuga Lake. The warping cultists correlate the 940-foot delta levels with the 900-foot Horesheads overflow level by ascribing the 40-foot difference in altitude to postglacial northward uplift. Actually the 940-foot level marks the altitude at which Lake Ithaca first drained westward across the upland separating the Cayuga and Seneca lakes basins in the vicinity of Ovid. As soon as the ice front had retreated northward far enough on this slope to expose land between the lakes at any level below 980 feet, Lake Ithaca began to drain down into Lake Watkins. At 940 feet either the ice front was stabilized for an appreciable interval or a relatively enduring rock barrier was encountered which temporarily fixed a lower level of Lake Ithaca. Besides the delta-top altitudes, the existence of this 940-foot level is confirmed by an overflow site and delta at North Lansing (N.Y.S. Route 34, one mile east on the

The High-Level Lakes

intersecting road at North Lansing). This overflow is exactly at the 940-foot elevation and brought drainage from the Owasco Lake basin to Lake Ithaca and built a delta precisely where it entered Lake Ithaca. (This delta has since been entirely removed for use as road gravel.)

During this 940-foot interval of Lake Ithaca the pattern of the ice front and the ice cover apparently was complicated. In part, this irregularity may have been due to the presence of the north-south Salmon Creek valley acting as a feeder to the ice current of the Cayuga Lake valley. Such irregularity is deduced from the circumstance that, eight miles south of the North Lansing 940-foot delta-top site, Route 34 crosses the line of transition from lake-clay deposit to unsorted glacial soil at exactly the 900-foot altitude. This line quite clearly marks the Lake Watkins overflow level at Horseheads—a correlation that is further confirmed by the circumstance that the lake-clay deposits are significantly thick, indicating the relatively long existence of Lake Newberry. This is the name given to the lake formed when Lake Ithaca and Lake Watkins were combined by open water across the upland between the Cayuga and Seneca lakes basins. Besides, delta tops were developed at Ludlowville at the 940-foot and 900-foot levels. But south from North Lansing there seem to have been areas to which the 940-foot lake waters were denied access because of covering ice. This ice probably consisted of inert slabs that persisted at least until the 900-foot level of Lake Newberry had been established.

As the ice front retreated, the extent of Lake Newberry was greatly increased. The lower lands at the northern ends of the Finger Lakes permitted the high-level waters of Owasco and Skaneatelas lakes on the east and of all the lakes west of the Seneca basin and even of the Genessee River valley lakes to be connected and combined to produce a greatly expanded Lake Newberry, which continued to drain into the Susquehanna through the Horseheads 900-foot overflow.

The Horseheads outlet was abandoned when the ice front had

retreated to the vicinity of Batavia, New York. There a westward overflow to the Mississippi drainage was opened at approximately 825 feet. This stage in the lake history is called Lake Hall.

There followed a series of westward overflows at successively lower levels as the ice front retreated. Eventually, at Rome, New York, at the elevation of 460 feet, an eastward escape to the Atlantic was opened up through the Mohawk and Hudson rivers. At this time Lake Iroquois, the high-level precursor of Lake Ontario, came into existence. But even at this late stage an arm of Lake Iroquois projected up the whole length of the present Cayuga

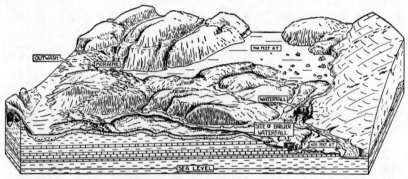

Fossil waterfall and high-level proglacial lake overflow at Clark Reservation, near Jamesville, New York. On the right is seen the front of the glacier ponding the water up to the base of the moraine at Tully, New York, in the Onondaga Valley. (Drawing by Steve Barker.)

Lake. It was only after the St. Lawrence River outlet was uncovered and Lake Ontario came into existence that the high-level Finger Lakes history came to a close and the present drainage was established.

Although the site is slightly outside the Finger Lakes Region, an account of Clark Reservation, west of Jamesville (N.Y.S. Route 173), is appropriately included at this point. The spectacular features of Clark Reservation were developed at a stage when the ponded waters of the region were afforded an escape eastward to the Mohawk River–Hudson River drainage at a 740-foot overflow level. Then, for a considerable period, the high-level Finger

The High-Level Lakes

Lakes were expanded to include the Onondaga Valley south of Syracuse. The initial Onondaga Lake overflowed into the Tioughnioga River drainage at the altitude of 1,200 feet. But a slight recession of the ice front brought about a merging of the original Onondaga Lake with those of the wide lake to the westward succeeding Lake Hall. At this juncture the ice front stood at the crest of the Onondaga limestone escarpment in the Jamesville area. This alignment permitted an overflow at the 740-foot level over the top surface of the limestone. At the Clark Reservation the waters plunged over the edge of the escarpment from the 740-foot level to the 420-foot level of the escape to the Mohawk River drainage. There was thus created a waterfall 320 feet high; over twice as high as the present-day Niagara Falls. Although the volume of the water that poured over the brink probably was not as great as that of Niagara, still it represented the drainage from precipitation of a wide area westward, and this was augmented by melt water from an equally wide part of the glacier front. All in all it must have been a magnificent spectacle, but there was not even an Indian around to see it, for it was operative at least 9,000 years ago. However, the "fossil waterfall" and the plunge pool created at its base, now occupied by a deep circular lake, are impressive records of the episode. As at Niagara, the retreat of the falls gave rise to a gorge which enhances the impressiveness of the site. Clark Reservation is well worth seeing merely as a scenic feature but the visitor who knows the circumstances of its origin enjoys as well the gratification that derives from a mind's-eye reconstruction of its functional grandeur.

A feature of the Fall Creek valley, a development related to the proglacial lake history of the region, and one that is, perhaps, unique in occurrence in this association, is the series of large incised meander curves that appear in the reach of the stream between the villages of Varna and Forest Home. (N.Y.S. Route 392). There, cut banks and terraces are evidence of a period during which this section of Fall Creek had a low gradient and, consequently, a meandering flow. The cut banks are 50 to 100 feet high, and the terraces at their base are 10 to 30 feet higher than the present stream bed, that is, far above the levels of the highest flood levels reached at present. The cut banks and terraces were eroded into unconsolidated glacial deposits and lake-bottom sediments. The altitude of the floors of the terraces is around 900 feet.

Accordingly it is clear that the meander erosion was done after the Lake Newberry level had been established. This 900-foot level may have provided the local base level that limited the depth of downcutting possible for Fall Creek in the period immediately following the draining off of high-level lake waters from this part of its valley. Since only unconsolidated materials were encountered down to the 900-foot altitude, the downward excavation of a narrow trench to this depth on a steep gradient probably took only a short time. Thereafter, and as long as the Lake Newberry level was maintained, the stream developed a low-gradient meandering flow, and the lateral swings carved the terraces and cut banks.

The decline in level from that of Lake Newberry to Lake Hall abruptly increased the gradient of what was then the section of the stream just above its outlet. But the downcutting this rapid flow induced was abruptly checked only a few feet below the 900-foot level by the stream's encounter with bedrock. Thereafter the ages-long program of cuesta development by downdip migration of the stream course and of escarpment retreat was resumed after an interruption of only the mere million years that comprise the glacial epoch. This process has lowered the stream bed postglacially only a few feet, as exemplified, at the Flat Rock site previously described.

A special interest attaches to the first terrace in Forest Home on the south side of the stream (immediately upstream from the bridge at the east end of Forest Home on N.Y.S. Route 392). This is an example of what are termed rock-defended terraces. At its eastern end bedrock crops out at the base of a spur that comes down to the level of the road. When the stream encountered the rock core of this spur on its upstream side, it was shunted over

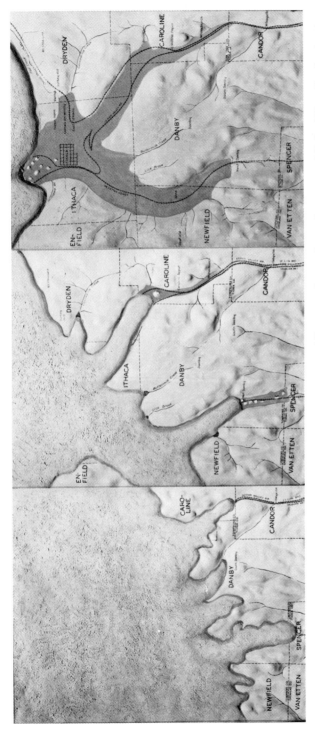

Proglacial lake succession. The model map on the left illustrates the conditions when the lobate fronts of the glacier were stabilized at points that are now the in-valley water partings between drainage to the Susquehanna and to the St. Lawrence rivers. The middle map shows the early stages of Lake West Danby in the Cayuga Inlet valley and of Lake Brookton in the Sixmile Creek valley. The map on the right shows Lake West Danby and Lake Brookton combined to form Lake Ithaca.

Drumlin, showing the characteristic shape of this kind of glacial deposit. The steeper left end is north. (Photographer not known.)

Moraine soil with angular boulders deposited by the glacier along North Triphammer Road at the altitude of 920 feet (above highest proglacial lake level at this location, hence no lake clay). (Photograph by J. D. Burfoot, Jr.)

The High-Level Lakes

to the north bank and came back to the south side only after the rock spur had been bypassed. The terrace (flower gardens) on the upstream side of the spur is especially large and well developed. That on the downstream side is smaller and less perfect, because the presence of the rock spur interfered with the free swing of the stream. Accordingly, the downstream terrace is said to be rock-defended.

The higher levels, overflow sites, and deltas of the Finger Lakes "West" have been little studied and afford interesting exploring possibilities to anyone inclined to research of this type. The new large-scale topographic maps permit close correlation of altitudes without recourse to the laborious field leveling formerly necessary for assured interpretations of the proglacial lake history of the Finger Lakes Region.

8 ★ *Glacial Deposits*

AT almost every turn in the preceding chapters there has been a direct or indirect reference to glacial deposits. Glacial deposits obscure the old escarpment lines, have disorganized much of the preglacial drainage, represent the enormous volume of rock excavated by the processes of glacial erosion, provide the minor relief of the landscape, and are the basic materials of the soils of the Finger Lakes Region.

In these respects glacial deposits can be regarded as incidental to the fact of the ice invasions. Beyond this, however, such deposits are themselves distinctive in origin, composition, and form, sufficiently so, in fact, as to comprise features of particular and peculiar interest to the discriminating observer of landscape aspects. The minor irregularities of the topography of the fields and pasture lands are not just chance variations. Instead each different item is the product of particular processes and circumstances. From these it derives its distinguishing characteristics.

There is, further, a marked degree of ordering in the distribution of the numerous varieties of deposits and considerable dif-

Glacial Deposits

ferences in their size and form. The distinctive aspects and the ordering provide the bases for identification and systematic description of the distinctive occurrences.

All glacial deposits now in evidence are the relics of the last phase of the ice occupation of the region. The deposits made by earlier ice advances were swept away by the last ice advance, except as some buried remnants may remain in protected nooks.

There is regularly present a basal layer of ground moraine, also referred to as till or boulder clay, but, only when it appears as a thin veneer over the bedrock of hill-summit tracts is ground moraine a landscape feature. Normally it is covered by other types of deposit and is exposed to view only where a road cut or other excavation has penetrated the overlying materials to below the level of the ground-moraine surface. In such a sectional display the "boulder clay" characteristics are strikingly exhibited. The face of the cut shows a boulder-studded clayey ground mass. This is regularly very compact. If recently exposed it is dark blue. Nearly all the boulders it contains are glacially striated and faceted. These boulders are mostly specimens of massive or durable types of rock—limestone, granite, gneiss, quartzite. Characteristic ones have been fashioned to a flatiron shape, that is, have a pointed nose, a flat-soled base, and a rough broad end at right angles to the nose. Boulder clay when encountered by farmers in tilling the hill lands or in digging wells is referred to as hardpan. (Outside of glaciated regions there are other types of hardpan.)

Boulder clay, or ground moraine, is deposited material transported at the sole of glaciers. It comprises the rock substance eroded from the bedrock over which the ice has flowed. Once ground off or torn loose, the material, while being carried along, is probably subject to intense kneading and attrition under the pressure of thousands of feet of overlying ice. In fact the bottom layers of a continental ice sheet, through a considerable thickness above the bedrock surface over which the glacier is moving, are probably a mixture of ice and rock debris. Although what goes on cannot be learned from actual observation, it may be inferred from the effects produced. Thus the clay matrix of the deposited ground moraine is largely made up of glacial rock flour, particles so fine as to be an impalpable powder. Such material derived from light-colored rocks, clouds the streams produced by the melting of glaciers in the Alps and has given rise to the term glacier milk, because the water strikingly resembles the lacteal fluid. The final deposit of the rock flour, studded with boulders, at the glacier front may be conceived as a large-scale troweling process, the plastering on of the ground moraine in successive layers as the material is brought forward by the ice. This manner of deposition gives rise to the notable compaction which is a marked characteristic of the boulder clay.

It was stated above that the over-all veneer of ground moraine has little or no topographic expression. On the summit tracts and upland slopes the deposit is commonly too thin to display independent relief. This generalization however, is, subject to one striking exception, the drumlin phenomenon. Although the drumlin belt of New York State is slightly outside, north, of the Finger Lakes Region, strictly defined it is so close an adjunct to it and so noteworthy a landscape feature of the general area that its inclusion in this account is quite warranted.

In brief, drumlins are hills composed of ground moraine. These hills are sufficiently distinctive in form and are so topographically prominent, so uniformly oriented, and so definitely grouped as to have earned recognition by the rural populace, which calls them hoddy-doddies. Hoddy-doddies range in length from a few hundred feet to a mile or more, in height from fifty feet to several hundred feet, and are from one-fourth to one-half as wide as they are long. Their long axis trends slightly toward the northwest from a north-south direction. Their north ends are steep and blunt; the south ends are gently sloping and taper off. Their sides are smooth curves. Drumlins are commonly likened in form to that of an egg bisected lengthwise and set down on the cut surface. Between drumlins the land is singularly level.

In New York State the drumlins occur in an east-west belt im-

Glacial Deposits

mediately north of the northern ends of the Finger Lakes "East." There are said to be nearly 10,000 of them.

The question as to how this extraordinary assemblage of uniquely similar topographic units came into being has been much debated. There are two possibilities. One is that a great thickness of previously deposited ground moraine was dissected by a minor readvance of the glacier. The other is that they were built up under the thin wedge of the terminal area of the ice at points where the transporting power of the glacier for its burden of ground moraine was slightly deficient and thus led to a concentration of deposit at those places. The latter concept is supported by significant evidence. Since the occurrence of the drumlins is staggered, it may be inferred that the distance between one drumlin and the one next in line with it served as a supply area for the upbuilding of the more southern unit. Again, rock knobs of similar form (though of much smaller size), prominent in the Thousand Islands of the St. Lawrence and in the Adirondacks, are unmistakably the products of ice erosion, although they reverse the drumlin orientation. Such rock knobs are tapered toward the ice front, massive and steep-cliffed at the other end, whereas the blunt, broad, steep ends of drumlins face the direction of ice advance. Further, the material composing drumlins appears plastered on while the surrounding substance of the rock forms was rubbed away. Finally, the occurrence of the New York State drumlin field is roughly just north of the line of the Onondaga escarpment so that it may be deduced that this topographic barrier adversely affected the free forward transport of the ground-moraine materials of the basal ice and induced their accumulation to compose the drumlin swarm.

It must not be inferred that all the debris of glacial erosion was transported at the sole of the glacier. Actually fragments great and small were incorporated in the mass of the glacier and, especially near the terminus, large amounts rode along on the surface of the ice. They got into the upper levels of the ice when the glacier flowed over mountains, such as the Adirondacks, or were moved from the base of the glacier by internal currents directed upward. If the volume of such englacial and superficial debris was great, it could provide a sheet of loose materials spread over the compact ground moraine when the ice over a given site finally melted. There was, however, no progressive accumulation. Only the material in and on the ice at that time could contribute to the cover. Where such deposited loose material is conspicuously in evidence, it is called ablation moraine. But even where the ablation moraine is relatively thin, it was, over wide areas, the material from which postglacial soils were formed.

Occasionally a huge boulder of crystalline rock, commonly derived from Adirondack sources, was left behind on the surface of the ablation moraine. These boulders are called erratics and probably rode along on the top of the ice to the place where they were let down. One such enormous erratic was brought in from the Sixmile Creek valley and carved to form a seat at the southwest corner of McGraw Hall on the Cornell University campus as a memorial to Professor R. S. Tarr who deciphered much of the glacial history of the Finger Lakes Region. Another may be seen in the bed of Fall Creek just downstream from the bridge at the east end of the village of Forest Home. A third example, perhaps the largest one in these parts, occurs on a hill summit immediately north of the village of Slaterville Springs.

Although the drumlin swarm, collectively considered, is of vast bulk, there is yet another accumulation, perhaps equally massive, of glacially derived rock waste. This is present in the ridges and hummocks of the region, which attained such prominent development that they have acquired distinctive designations. They are now referred to as "Valley Heads Moraine"; at an earlier time they were identified as "moraines of the Second Glacial Epoch." They are called terminal moraines, end moraines, and recessional moraines.

The last glacier invasion extended as far south as Williamsport, Pennsylvania, in the meridian of Seneca Lake. Over the southern 75 to 100 miles back from this terminal line the ice occupation appears to have been of short duration and feeble action. From

Glacial Deposits

the most southern position there was a rapid wasting back of the ice front to a line with a northeast-southwest trend, situated, in general, a short distance south of the heads of the Finger Lakes. At this line the ice front was maintained for a prolonged period in an equilibrium between supply and melting. Estimates range from 200 to 500 years. During that time all the rock debris that the ice was transporting at its base, in its interior, and on its surface was deposited at the ice front and now appears as ridges and hummocks, which in places have impressive dimensions. At the few sites where ridges have been completely sectioned by postglacial stream erosion (e.g., Fall Creek, east of Varna), they are found to have a core of compact, dense boulder clay—the ground moraine of basal transport. The upper parts are made up of a heterogeneous mixture of looser materials of all sizes ranging from huge boulders to fine clay with lens-shaped layers of water-sorted stratified materials. This jumble is the debris dumped from the surface and the interior of the ice at the end of the glacier. The hummocky topography of the moraine is indicative in part of the inequality of supply along different lines of the ice current, in part of contemporaneous burial of stagnant ice blocks which, when they eventually melted, caused the covering material to subside and gave rise to a kettle, either dry or filled with water—a kettle pond. In general such a moraine is not a single ridge; rather it is a belt of ridges since such perfection of equilibrium in supply and wasting as to give rise to a single ridge would be improbable. Each ridge of the belt, however, represents the last time the front had come that far forward.

There is some development of end-moraine ridges in practically every valley of the region, but little on the slopes of the intervening rock hills. As previously noted, the ice lingered longer, because thicker, in the north-south valleys, which also were the lines of active flow. The front of the ice, accordingly, was a pattern of projecting tongues which closely resembled in form and function the valley glaciers of mountains. Since the axis line of a valley is where it is deepest, the tongues extended farthest down the center of the valley. In consequence the moraine deposit at the end of a tongue was a crescentic loop, convex down-glacier, extending from valley side to valley side—a morainic loop.

I have seen a number of glaciers and photographs of many more, but only a few are building sizeable end moraines. This may be because glaciers at present are shrinking, whereas the accumulation of a bulky end moraine would require the maintenance of a stable ice front for a considerable period. There are indications that end moraines resulted when the equilibrium period was preceded by a significant forward thrust of the ice following a period of slow retreat. Such a pulsation could result in a dumping of debris from the upper surface of the ice and a shoving ahead of previously deposited moraine.

Perhaps the largest and most representative example of the morainic-loop phenomenon in the Finger Lakes Region is that in the Cayuga Inlet valley at North Spencer (N.Y.S. Route 34-96). A climb up its steep front and a walk across its width at the summit will repay the reader who seeks in imagination to restore the vast ice mass which filled the deep, wide valley basin and rose in the distance to the north so as to overtop the highest hills. If then, further, he develops a mental picture of the morainic debris rolling, sliding, and washing down the end of the ice tongue and piling up irregularly at its front, in places burying the ice itself—in sum, building a hummocky surface of knobs and kettles littered with boulders—he will have achieved a real awareness of conditions along the glacier front in the Finger Lakes Region at the time in the Ice Age when the end moraines were being built.

In some respects even more impressive than the Inlet Valley moraine at North Spencer is that closing the end of the Onondaga Valley at Tully (U.S. Route 11). While the Inlet Valley development is more imposing when viewed from the south, the Tully moraine is especially convincing as seen from the north. Further, at Tully, a commanding overlook of the deposit is had from the road on either side of the valley. In this area there is also present an unusual feature, a notable morainic accumulation where the

Glacial Deposits

ice front crossed over from the Onondaga Valley to the parallel, next east, valley of Buttermilk Creek. At the crest of the hill on an unnumbered road leading directly north from the village of Tully the landscape is a perfect example of end-moraine topography, a wide panorama of knob and kettle forms. It seems that what could be regarded as overflows of the main ice currents in the two closely adjacent valleys provided sufficiently active forward flow of the glacier in the divide area between the valleys to bring about the transport and deposit of this notable accumulation of debris. At such sectors of the glacier front ice movement is inferred to have been very sluggish.

At Groton, in the Owasco Lake valley, a sharply defined end moraine has its ridge crest right in the center of town so that the main street ascends over its hump (N.Y.S. Route 38).

Particularly intriguing end-moraine developments are encountered where ice lobes protruded into east-west valleys as branches from the major north-south currents. These were the final expression of relatively strong basal ice flow through these lateral valleys at the climax of glacierization. Thus, in the Fall Creek valley such a lobe, projected from the Cayuga Lake valley, was maintained for a long period at the close of the glacierization with its front at Varna (N.Y.S. Route 13). The morainic mass then accumulated is ascended from the west at the Varna hill. The crescentic moraine-loop concept of the deposit is convincingly presented by the view from the base of the hill.

This phenomenon of lateral lobes as recorded by end moraines has its most interesting demonstration where east-west-trending valleys extended more or less continuously between the major Cayuga and Seneca lakes north-south channels of ice flow. At the time of the long halt in the recession of the glacier, which permitted the building up of the massive end moraines, such arms in two instances had wasted down just so far as to have separated into two tongues ending only a few miles apart. In the first instance the lobes extended up the Trumansburg Creek valley from the Cayuga Lake valley and the Hector Creek valley from the

Seneca Lake valley. The Cayuga lobe was evidently dominant because it built its moraine at Reynoldsville (N.Y.S. Route 227) four or five times farther from the Cayuga Valley than the moraine at Bennettsburg of the lobe fed from the Seneca Valley. At the crest of the ice flood of the first advance of the glacier, the two lobes merged and their combined volume moved southward through a preglacial stream valley tributary to the Reynoldsville-Bennettsburg channel. Eventually the ice flow reamed out this preglacial tributary valley to the great dimensions of the Texas Hollow trough.

The other notable development of the same kind occurs in the Pony Hollow valley about four miles west of the village of Newfield (N.Y.S. Route 13). In this instance the thrust of ice from the Cayuga and Seneca valleys was more nearly equal; the opposed moraines are at about the same distance removed from the two major, north-south feeding currents. The Pony Hollow moraines, however, are not so conspicuously opposed as those at Reynoldsville. But once attention is called to the situation the significance of the deposits is easily grasped. The exact site of the one-time contact of the two lobes is marked by a quite characteristic overflow channel which extends across the road to the south side of the valley. This channel appears to have been the outlet of a small lake ponded on the north side of the lobes.

Although numerous references have been made previously to waters originating from ice melting, no attention has been given to their significance other than to note their serving to fill proglacial lake basins to overflow levels. The existence of most such basins implies that the front of the glacier was present as a barrier beyond a land divide between north and south drainage. It has been pointed out, however, that at the sites of the two greatest morainic loops, those in the Cayuga Inlet valley and the Onondaga Valley at Tully, the moraines mark the postglacial land divide. Accordingly, the water released at the moraine front had in these instances a downgrade escape to the Susquehanna drainage.

Glacial Deposits

The overflow waters of large lakes are clear and of nearly uniform volume. They enlarge their outlet channels by the sweep of their currents. These characteristics are most magnificently exemplified postglacially by the upper Niagara River. The nature of waters that emerge from and drain directly off a glacier front is altogether different. Whether the glacier front terminates in water or on land, there is first deposited under the ice the ubiquitous sheet of boulder clay. But much boulder clay is carried forward to the ice front by subglacial streams. Where these escape under water their whole load of coarse debris is dropped at once and provides the material for the upbuilding of sublacustrine moraine accumulations. In these circumstances there is, however, an immediate sorting process. The rock flour of glacial grinding remains in suspension for considerable distances beyond the glacier front. Eventually it settles out and sinks to the bottom of the lake as glacial lake clay, and the presence of this as a surface soil clearly marks the former existence of a glacial lake over the area. If the ponded glacial lake was shallow, as was not uncommonly the case, the volume of the sublacustrine moraine deposit and of the lake-clay cover often sufficed to fill the basin entirely. Then the situation was the same as that obtaining where there was a downgrade of the land from the glacier end.

Melt waters, as they come out from under the ice, flow along the margin of a glacier tongue, or trickle down the upper surface of the ice are burdened to capacity with the rock debris which is the product of glacier erosion. This is especially true of subglacial streams. While under the ice they flow with great velocity and force under a hydraulic head. When such streams emerge from under the ice their power is suddenly spent and the coarse boulders (maybe a foot or two in diameter) and pebbles they were transporting are dropped immediately. Streams marginal to the ice lobes are hardly less competent in the transport of debris. They are regularly roaring torrents because of the steep gradient provided by the downslope of the terminal portion of the ice tongue. The water from the glacier surface is muddied by the fine debris incorporated in the ice, the coarser stuff remains behind and, like the ground moraine, is deposited, as ablation moraine, when all the ice finally melts.

Although a great amount of the rock debris of glacial erosion is deposited by the ice directly, such accumulations account for only a part, and it may be only a small part, of the vast abrasion and excavation wrought by the continental glaciers of the Ice Age. The floods of melt water that poured downgrade from the ice front were torrents of mingled water, boulders, grit, sand, and clay. Wading hip-deep across such a stream emerging from the front of the Alaskan glacier, which most closely simulates the conditions existing at the margins of the continental glaciers of the Ice Age in Central New York, at the farthest point upstream where this could be ventured, was dangerous because of the force of the current and because the footing was a moving sheet of boulders several inches or more in diameter. The harsh clash of these bottom boulders in transport is clearly audible above the noise of the rushing waters to an observer posted on the side of the stream.

If it is kept in mind that such melt-water conveyance of debris was operative not only at the close of the Ice Age but also throughout this epoch, it will be appreciated that tremendous volumes of rock waste were hurried down to the sea during that time. At the margin of a glacier some of the large boulders are immediately deposited at the ice front. Others are worn down by attrition in the mill operating on the stream bed, first to pebbles, then to sand, and thus are converted to fine materials which, together with the rock flour of glacier grinding, can be carried by the gentler currents of the downstream reaches.

Although the over-all transport of the rock waste by the outflow streams is thus voluminous, there is at the same time a notable building up, or aggradation, of the stream bed. Wherever and whenever the load of debris is too great to be moved forward by the velocity of the current, part of it is left behind. In consequence of such deposition, at the close of the glacial epoch vast accumulations

Glacial Deposits

of gravel and sand, called outwash plains, were made on the floors of the south-draining valleys.

Outwash plains are not scenically spectacular. They are, however, distinctive landscape features in the Finger Lakes Region, in that, with the exception of the postglacial deltaic accumulations at the heads of the lakes, they are the only widespread, low-altitude, level lands that occur there.

There are a number of noteworthy developments of the outwash-plain phenomenon within the area. In the east of the region, south of the Tully end moraine, the Tully-Cortland plain, filling the Tioughnioga River valley, is impressive both as to its surface extent and its perfection of development. Its level surface is ideal for cultivation. Further, the soil is highly productive. When a given line across the valley was left behind by the ice front as it melted back toward the north, the coarse boulders and gravel that had been deposited at the immediate margin were covered with the fine sands and silts laid down later when the velocity of flow there had declined. These fine-textured materials provide excellent soils for present-day agriculture.

A special feature of the Tully-Cortland plain is the numerous shallow lakes dotting its surface. These are irregular in outline and range in size from mere ponds to considerable expanses of water. Because of them this exhibit qualifies for the designation "pitted outwash plain." By analogy with the same phenomenon at the fronts of existing Alaskan glaciers, the Tully lakes are quite certainly sites where stagnant ice blocks were rapidly buried by the outwash gravels. Long thereafter, when deposition had ceased, the buried ice blocks melted away and the material overlying them slumped down to form the pits, or basins, of the present-day lakes.

The fact that the pits are water-filled instead of being dry kettles is evidence of saturated soil at only a shallow depth below the surface of the plain. An interesting consequence of this ground-water condition is that it permits the large-scale recovery of gravel for road and concrete construction by dredging. To the east of the road

(U.S. 11) between Homer and Tully this is being done on a large scale. The top four or five feet of fine-textured soil is removed, the pit fills with ground water, in this a dredge is floated and proceeds to scoop out the unweathered coarse gravel which is encountered at the lower levels.

At the southern end of the Tully-Cortland plain, west of Cortland, the retreat of the glacier front in the Tioughnioga, Otisco, and Skaneateles valleys was for a period so rapid that the coarse materials laid down close to the ice margin did not get completely buried by finer materials. Hence the soil in this area is a mixture of small boulders and sandy silt. This, rather surprisingly, has been found admirably suited for growing beans. It seems that the boulders store sun heat and promote drainage so that a warm dry soil results which exactly fits the needs of the bean plant.

Another large outwash plain is in the Cayuga through valley south of the end moraine at North Spencer. A third example is that around Alpine (N.Y.S. Route 224) in the Pony Hollow valley.

All three of these are, however, really small in comparison with the widespread outwash plain south of the end moraine in the Seneca Valley trough. All the valley floors between Pine Valley (N.Y.S. Route 14) and Horseheads, between Horseheads and Corning (N.Y.S. Route 17), from Horseheads through Elmira (N.Y.S. Route 13) are one uninterrupted outwash plain. As in the Tully-Cortland occurrence these outwash plains also are exceptionally good agricultural acres—level, easily cultivated, with deep, well-drained, productive soils situated in the midst of the hill lands of the southern sector of the Finger Lakes Region, which are for the most part suited only for pasture.

To round out the listing of outwash plains, mention needs to be made of the area between Wayland (intersection N.Y.S. Routes 21 and 63) and North Cohocton. There are still other lesser developments which might be regarded as outwash plains but really belong better with what are called valley trains. Every valley sloping southward to the Susquehanna of necessity served as a channel for the melt waters of the glacier. Except as these waters became lake over-

Glacial Deposits

flows, they were loaded to capacity with sediment. With reduction in velocity downstream their transporting capacity was reduced; the sediment was progressively deposited and built up a valley "train" (string-out) on the floor of the valley. Actually valley-train material extends all the way down the Susquehanna to the ocean. In the lower course of the river the only identifiable glacial materials that remain are pebbles of the very hardest rocks, such as quartzites and granite. Less-resistant types have been ground down to fine sand or clay particles in the course of the trip after their emergence at the glacier front.

Representative examples of valley trains are the floors of the Tioughnioga River south of Cortland; Owego Creek, south of Dryden; Cayuta Creek, south of Van Etten; Cohocton River, Cohocton to Corning; Canisteo River, Hornell to Addison.

As has been noted, outwash plains and valley trains, because of their low relief, are not scenically impressive. On the other hand, two topographic features—kames and eskers—with attention-arresting characteristics owe their existence to particular concentrations of glacier melt water.

Both kames and eskers are glacially conditioned deposits composed exclusively of stratified gravel, hence clearly owe their construction to the sorting action of moving water. Kames, typically, are extremely hummocky accumulations, highly exaggerated expressions of the irregularities commonly marking the surface relief of end-moraine ridges. Close-spaced knobs and kettles, the latter up to fifty and more feet deep, but dry because of the porous gravel composition, comprise the topography of a kame deposit. The deposits, as a whole, are massive and in outline irregular. From this description it will be appreciated that a kame area is a rather distinctive aspect of landscape.

Aside from general agreement that kames, like end moraines, are built up in immediate association with the front of a glacier, explanations of the circumstances of kame accumulations are diverse. Kames are evidently built up where the glacier is quite stagnant for some distance back from the front. Then large volumes of melt water are collected in subglacial channels and flow through these under hydrostatic pressure. At the ice front such streams burst forth from underneath the ice and literally spew out their enormous sediment load of sand and boulders. Thus the thin edge of the stagnant ice tends to get buried under a thick cover of this debris. Eventually, perhaps long after the ice front has receded from that line, the buried ice melts.

In the Finger Lakes Region such stagnant ice wedges appear to have developed where, in the course of the retreat of the front, there was a wide low area at the southern end of a strong, confined, feeder-current of ice, such as one of the Finger Lakes troughs, and where the far side of the low area was bounded by a hill barrier with only constricted gaps across it. The glacier then spread out feebly over the low area and banked its edge against the barrier hill slope. In effect, the materials which would elsewhere be piled up as an end moraine are in these circumstances water transported to build a kame area.

A conspicuous kame deposit which fits these specifications is on the south side of N.Y.S. Route 13 between South Cortland and Cortland. It can be identified as such because of the gravel pit dug in the side of one of the knobs. Its location conforms rather exactly with the type of situation outlined above as conducive to kame deposition. The wide valley floor at Cortland brought about the lateral expansion and incipient stagnation of the strong ice current emerging from the broad Tully-Cortland trough. The outer edge of this fan-shaped wedge of ice was banked against the hill slope south of Cortland. Escape southward could only be through the constricted Tioughnioga River valley, entrance to which is offset several miles to the east.

An even more representative kame area occurs near Freeville. This development is at an "off the main route" site. Leave N.Y.S. Route 13 north of Dryden at the intersection of N.Y.S. Route 38 and follow the first road (hard surface but unnumbered) to the right (north) over the hill to its dead-end intersection with a dirt

Glacial boulder near Noyes Lodge, Cornell University. A block of Onondaga limestone brought by the ice from the outcrop of that formation some 30 miles to the north and imbedded in lake clay at the site of Balch Halls, it was uncovered during excavations for the buildings and preserved at the request of the author. Originally a sharp rectangular block, it was shaped and smoothed in transport under the ice, then floated by an iceberg to the site of its deposit. (Photograph by J. D. Burfoot, Jr.)

Wasting glacier in upland country. This picture of a terminal section of the Hardanger Glacier of Norway may be regarded as representative of the conditions in the upland sections of the Finger Lakes Region when the glacier was melting off the countryside. (Photograph by O. D. von Engeln.)

Moraine topography, Cayuga Inlet valley, looking south. The end-moraine ridge which marks the in-valley divide between the Susquehanna River and the St. Lawrence River drainage shows faintly in the upper right-hand corner of the picture. The hummocky valley floor is morainic deposits smoothed over by deposits of lake clay, for this was the bottom of proglacial Lake West Danby. A section of the valley side oversteepened by glacial erosion is well shown on the left side of the picture just above the middle distance. The upper left corner shows the topography of the peneplain uplands of the Finger Lakes Region. (Photograph by C. S. Robinson.)

Glacial Deposits

road. A short distance beyond a pasture lot immediately north of this intersection one can see perfect kame topography characterized by extremely deep, dry kettles. The kame tract extends eastward and northeastward for several miles. In dry weather the dirt road may be followed eastward to its intersection with Mott Road, where there is a large kettle several hundred feet deep.

As in the case of the Cortland occurrence this kame tract resulted from the lateral expansion of a strong ice current emerging on a broad lowland, the Fall Creek valley floor, from the Owasco Lake valley trough and banking against the hills to the south. Here, again, the escape route is devious, through the Dryden–Owego Creek valley.

This Freeville kame tract is especially interesting because it is linked up directly with a notable esker development, perhaps the best example of this phenomenon in the Finger Lakes Region. In previous paragraphs the collection and flow of glacial melt waters in subglacial channels was outlined. If this occurs where the ice is actively moving, all trace of such channels disappears when the ice finally melts off, presumably because the lines of flow shift as the currents vary. But when a slab of ice becomes stagnant and inert, a subglacial line of flow, once established, can be maintained until the ice finally melts completely.

These conditions are also ideal for kame construction. Where the subglacial stream emerges at a site where there is no downgrade to permit the further transportation of the glacial debris it carries or to permit the construction of outwash plains and valley trains, there must be immediate deposition at the outlet of the tunnel. Such blocking of the flow compels a slackening of the velocity of the current under the ice and the deposition of some of the debris load on the floor of its subglacial course. The stream bed is thus raised to a higher level, which leads to the erosion of the ice roof of the cave. Maintenance of these relations at Freeville through the period required for the melting of the inert ice slab resulted in the accumulation of both a massive kame deposit and a narrow, sinuous, but thick deposit of gravel marking the course of the subglacial stream. Following the disappearance of the ice, this deposit remains as a sharply defined ridge—an esker.

The esker course that supplied the Freeville kame had its source north of the Fall Creek valley about one mile west of the village of Malloryville. Starting at Freeville, it is two miles on the Creek Road (not numbered) to Malloryville. At Malloryville a sharp left turn on an unimproved road (known as the Old Salt Road) leads first across Fall Creek, then across a railroad track to a stretch of road at the west base of the esker ridge. About one-half mile along this is a farm track leading northeast toward the summit of the ridge. If this is climbed where the farm track intersects it a fine prospect of the esker development is had.

The course of the esker across the level floor of the Fall Creek valley to the Freeville kame deposit has been erased, presumably by the postglacial flood-stage erosion of the stream. But a most convincing link between the esker and the kame accumulation is afforded by the materials in the gravel pit just beyond the entrance to the kame area at the site of the dead-end road. The gravel exposed there is studded with large, perfectly rounded, water-worn boulders. These boulders are almost exclusively of extremely hard or massive-rock types—quartzite, gneisses, granite, and dense limestone. Further, they are of kinds that could only have been gathered by the glacier from remote sources, the Adirondacks or Quebec, Canada. Anyone who wishes to make a collection of diverse, hard-rock specimens will find this a most rewarding source. The conclusion to be drawn from the nature and remote origin of these boulders is that only the most resistant types could survive the churning, grinding journey through the subglacial course to the deposit site. A confirmation of this inference is afforded also by esker material in another gravel pit dug in the ridge just beyond the railroad crossing. There the boulders are much less rounded and water-worn, and many of them are derived from local-rock formations. In other words, the longer the subglacial trip the more complete was the elimination of the soft local materials and the

Glacial Deposits

concentration of the relatively scarce resistant-rock types from remote sources.

There are other esker developments in the region but the Malloryville occurrence here described has almost ideal expression. A bigger one, locally celebrated, is called the Centerlisle esker. It occupies the valley of Dudley Creek on N.Y.S. Route 79 between Richford and Lisle. Although impressive because of its height and length, and because it is a ridge in the axis of the valley floor, it is a less-convincing example of the phenomenon because the circumstances of its origin are not so clear-cut as those of the Malloryville occurrence. The Centerlisle esker resulted from the stagnation of a tongue of ice that was projected eastward through a gap in the side of the Owego Creek valley at Richford. This was for a time a distributary line of ice flow but with the decline of glacierization was no longer adequately fed from the north-south valley, hence stagnated while still serving as a route for discharge of melt water from the trunk ice current and from its own inert mass.

To the reader who has persisted to these closing pages (or perhaps to the type of reader that refers first to the end of a book to learn how it turns out before committing himself further) it may appear as an anticlimax to have ended this account of Finger Lakes Region scenery with an elaborate description and interpretation of puny gravel heaps. Such reaction is not completely unwarranted since the concern of preceding chapters has been with impressive cliffs, picturesque lakes, romantic gorges, and rushing waterfalls, distinctly eye-catching and attention-arresting landscape features. The point of the book, however, is that these several scenic phenomena, whether of imposing or of minor dimensions, derive much of their interest from an understanding of the circumstances of their origin. With this approach the history of a kame tract and an associated slender esker is no less intriguing than that of a deep lake basin. Indeed, that the expression of such lesser items can be fitted into the larger picture is strong supporting evidence of the over-all faithfulness of the interpretations herein set forth.

As a whole the book should serve to demonstrate that to a perceptive observer evidence is available for compiling the history of a landscape that was many millions of years in the making. Thus the scenery of the Finger Lakes Region is revealed to have had, overall, an orderly, sequential evolution but in what was only the yesterday of its long geological history was drastically altered in many aspects by, and because of, the ponderous assault of the continental glaciers of the Pleistocene Ice Age.

Epilogue

AS is set forth at length in the preceding pages, the unique scenic features of the Finger Lakes Region owe their existence to the invasion and occupation of the area by ice sheets of continental dimensions during the Pleistocene glacial epoch. Once the fact of such glacierization was indisputably established, the question of what caused it was immediately raised. No conclusive answer to this problem has, as yet, been formulated. But the search for a solution has directed attention to the critical factors that governed the occurrence of the frigid visitation. Hence it is deemed appropriate to close this book with a résumé of the circumstances affecting the initiation of the Ice Age and of the hypotheses offered to account for it.

Glacial epochs are neither frequently recurring nor regularly spaced episodes in the history of the earth. The rock sequence which provides us with a record of several billions of years of geologic history affords evidence of only four such occurrences at widely separated time intervals and these with no common background except, perhaps, association with periods of crustal unrest.

In other words, the Pleistocene glacial epoch was distinctly in the nature of a departure from the normal course of earth history.

Although this cannot be asserted in regard to the three other known glacial epochs, it is quite definitely established that that of the Pleistocene was of world-wide incidence. The northern and the southern hemispheres and the equatorial latitudes were all affected and in comparable measure. This over-all occurrence rules out hypotheses of cause that are based on particular regional circumstances.

Third, the measure of cold essential for the development of continental glaciers in the middle latitudes, both as estimated on theoretical grounds and as deduced from actual evidence of the climatic conditions that then obtained, was a decline of only 15° to 20° F. from the present average annual temperature. No bitter polar cold was necessary to initiate or sustain the ice sheets, nor did such cold prevail.

Ample precipitation in the form of snow was essential. To start and maintain glacierization the volume of snow received in the cold months must appreciably exceed the amount melted in the warm season of the year. It has been demonstrated that the pattern of the planetary wind belts during the Ice Age was the same as that now in effect. Under the temperature regime specified above such atmospheric circulation could well provide the needed measure of snowfall.

The operational basis of this series of prescriptions is obviously the necessary degree of refrigeration. Except for the conduction of heat from the interior of the earth to its surface (of no significance in this connection) the radiant energy of the sun is the source of the warmth of the air, lands, and oceans. It follows directly that any considerable diminution of the sun's power will result in lower air temperatures, and from such a decline a glacial epoch could be initiated. This obvious relationship was of course appreciated early. It has, further, been demonstrated that solar radiation does fluctuate to a slight degree above and below its average level. But despite the relatively small decline in temperature required to pro-

Epilogue

duce an ice age the reduction in radiant energy necessary is far greater than any observed fluctuation even if this lowest level were to be maintained for a very long period. If, nevertheless, a sufficient failure of sun heat is arbitrarily postulated as the cause of glacial periods, the solution thus provided merely substitutes the problem of the cause of the solar defection for that of the cause of an ice age. Not so much as a purely speculative hypothesis has been advanced to account for so large a solar failure. But the incidence of glacierization from lower radiation fits the other specifications of unpredictable recurrence, of world-wide application, and of permitting ample snowfall with the known pattern of atmospheric circulation. Accordingly, despite the confession of failure to arrive at the ultimate cause many students of glacial phenomena are disposed to be content with the sun-failure postulation as an escape from the dilemma.

Nevertheless, the very moderate degree of refrigeration indicated to be all that was needed to bring about an ice age has prompted a continuing search for known or explicable astronomical or terrestrial factors that, suitably modified or correlated, would suffice to produce this. The slightness of the departure from existing temperature levels will be appreciated when reference is made to the figures applicable in the Finger Lakes Region. The average annual air temperature of the lower lands of the region, 400 to 800 feet in altitude, is 50° to 47° F. A decline of 15° to 20° F. would hardly suffice to initiate or maintain glacierization at those levels. But since there is a decline of 1° F. in average annual temperature for every 300 feet of rise in altitude, it appears that the summit areas of the region, at elevations of 1,800 to 2,200 feet, have an average annual temperature of 45° to 42° F. Subtracting 15° to 20° F. from these temperatures would give an average cold sufficiently below the 32° F. freezing point to support glaciers. In fact, during the the closing centuries of the Ice Age small independent glaciers persisted in the higher ranges of the Catskills (Slide Mountain, altitude 4,204 feet) after the surrounding regions were free of ice.

Perhaps the earliest hypothesis (not tied to the assumption of fluctuations in solar radiation) of the cause of the Ice Age to win considerable acclaim from persons competent to ponder its merits was that by Croll based on astronomical data. Croll pointed out that when the earth in its orbit is at one end of the long axis of the ellipse, it is much nearer the sun (perihelion) than when at the other end (aphelion). Further, that the earth's position in the trace of the orbit shifts so that if the northern hemisphere, owing to the inclination of its axis, is turned toward the sun in perihelion, 13,000 years later it will be turned away from the sun. Thus perihelion, northern-hemisphere summers would be short and relatively hot, the contemporary aphelion winters long and cold. Croll argued that the hot but short perihelion summers would not offset the cold and snow accumulation of the aphelion winters and that an ice age would ensue.

Although the approach is factually warranted, it was found by computation that the difference in perihelion and aphelion distances could not affect solar radiation sufficiently to induce the necessary refrigeration. Further, ice ages would need to alternate between the northern and southern hemispheres at short intervals and be recurrent throughout geologic time. There is no evidence that either the alternation or recurrence took place. The hypothesis, however, still merits attention because it suggests that increased eccentricity of the orbit, if coincident with terrestrial factors conducive to glacierization, might account for the necessary intensity of cold and explain the noncyclical occurrence.

Hypotheses that seek to find the cause of the Ice Age in terrestrial factors, solely, get deserved consideration because they rely on known phenomena. One such that has recently been well publicized is keyed to the idea that the sea-ice cover of the Arctic Ocean so greatly inhibits snowfall in the arctic polar areas that glaciers fail for lack of nourishment. The absence or presence of the ice cover is in turn attributed to the ability and failure, respectively, of warm water from the Atlantic to gain access to the Arctic Basin. This could result from the lowering of the sea level or the raising

Epilogue

of submarine barriers at the entrances to the north polar sea. It is, however, immediately clear that the rearrangement of ocean currents that would result, and the consequent change in the distribution of heat, could only affect the Northern Hemisphere. The concurrent world-wide expansion of glaciers in the Ice Age is quite unaccounted for.

The difficulty interposed by the necessity of providing for worldwide refrigeration in accounting for the Pleistocene glacial epoch is not operative in several other terrestrially predicated hypotheses. Thus it has been argued that the great volcanic activity of the geologic time immediately preceding the Ice Age so filled the atmosphere with erupted dust as to impede solar radiation sufficiently to induce the temperature decline necessary for the glacierization. It is, however, rather unlikely that such dust could be buoyed up for a long enough time to build up continental glaciers. Besides, it is awkward to account for two glacial advances within the glacial epoch on these grounds because there is no evidence of a repetition of intense volcanic activity during the Pleistocene.

Somewhat similarly, it has been proposed that fluctuations in the per cent of carbon dioxide and water vapor in the atmosphere could induce glacierization by reducing the radiant energy transmitted to the earth's surface. Although plausible terrestrial causes for such variations can be cited, it has been computed that the maximal content possible of the two gases in the mixture that is air would be quite inadequate to bring about the necessary degree of refrigeration, inconsiderable though this be.

American glacialists recognize both the limiting conditions that must be satisfied by an adequate explanation of the cause of the Pleistocene glacial epoch and the several defects, as outlined above, of the hypotheses that have been proposed. Near the close of a recent comprehensive review of the problem it is pointed out that glaciers today are regularly associated with highlands and that the highlands of the Western Cordillera and of the Labrador Plateau in North America, of Scandinavia and the Alps in Europe, of Patagonia in South America, were the focal areas from which the Ice

Age glaciers spread across the continental lowlands. This is followed by a listing at length of the world-wide increase in mountain heights that took place in immediately preglacial times. The implication is that high altitudes initiate glacierization. Accordingly, given sufficiently great altitudes continental glaciers could have a terrestrial origin.

This is entirely sound reasoning and exposition. Since mountains and plateaus, as such, are, however, apparently no lower now than they were in the Ice Age and since it has been demonstrated that the planetary wind belts had essentially the same distribution and characteristics then as now, the basic factor of the glacierization, namely the cause of the refrigeration, is left entirely unaccounted for. The pertinent question then is: Why should continental glacierization not have persisted into the present?

At this impasse the reviewer had no better recourse than to fall back on the speculation that a significant diminution of the radiant energy of the sun occasioned the necessary refrigeration. If such a decline is accepted as a reality, the build-up, leading presumably to an explanation based on terrestrial factors, becomes pointless. It is generally conceded that a sufficient defect in solar radiation could cause an ice age on earth.

Here is encountered one of those curious blind spots that interfere with the progress of science both pure and applied. (Thus Kettering's invention of the automobile self-starter depended on his seeing through a block set up by established electric theory.) Because of unwillingness to abandon preconceptions of the possible and impossible, obvious solutions are rejected and absurd hypotheses substituted to account for challenging phenomena. In relation to the problem of the cause of the Pleistocene glacial epoch the ignored, misunderstood phenomenon is the occurrence of submarine canyons.

The margins of the American continents and those of Europe and Africa are indented by submarine canyons of extraordinary dimensions: trenches that are miles long and 5,000 and more feet deep. By detailed soundings and diving explorations to consider-

Epilogue

able depths it has been demonstrated that the characteristics of these canyons match in every respect those of river-cut gorges of subaerial origin. Further, many of the canyons are seaward extensions of the valleys of existing rivers, examples are those at the mouths of the Hudson and Congo rivers. In at least one instance submarine strata laid down as recently as immediately preglacial geologic time have been cut through by such canyon erosion. Accordingly, the submarine canyons are not relics of some long-ago geologic episode; rather they are a record of cutting done during the Ice Age.

In sum, all the evidence points to the carving of the canyons by streams flowing overland. But for this to be possible it must be postulated either that the margins of the continents were 5,000 or more feet higher than now, with reference to sea level, or that sea level was that much lower than now, or that both higher lands and lower sea level prevailed in the Pleistocene glacial epoch.

(The circumstances attendant upon the erosion of the submarine canyons are exactly parallel to those which led to the excavation of the major gorges of the Finger Lakes Region. The opening of north-south through valleys across the region by glacial erosion introduced a steep, high slope at the ends of east-west valleys tributary to the overdeepened north-south troughs. When the ice melted off and the high-level lakes subsided, the tributary streams plunged down the steep slope in the same manner that the streams that carved the submarine canyons rushed down the steep slope leading from the margins of the uplifted continents to the deep sea floor. Although the Finger Lakes gorges are only miniature replicas of the submarine canyons, their characteristics completely match those of their great underwater counterparts. Buttermilk Gorge as seen by an observer stationed high on West Hill, Ithaca, so that his line of sight is an extension of the axis of the gorge is a most convincing exhibit in illustration of the above interpretation of the submarine canyon phenomenon.)

It is in regard to this concept of higher lands and lowered sea level that the closed minds are encountered. Glacialists who have no difficulty in postulating great variations in the energy of solar radiation, with no basis other than speculation, shy completely away from the idea of raising the continents en bloc, or lowering the bottoms of the oceans, or of such uplift and depression in conjunction. This rejection evidently derives from two circumstances. One is unwillingness to concede the possibility of world-wide changes in the altitude of the continental lands in relation to sea level. The other is the false inference that if such a shift had occurred it would have left a record inland in the form of canyons comparable in size to the submarine developments.

The first resistance may be countered objectively by pointing out that remainders of such continental upheaving persist in the level highlands of Africa south of the Sahara, in the lofty Venezuelan plateau and, less distinctively, in the Tibetan "roof of the world." Rivers now tumble in rapids and waterfalls and through gorges over the edges of these uplands in exactly the same manner that streams evidently poured over the sides of the continents in the Ice Age.

The second objection originates in a misconception of the status of the interior streams during this time of uplift.

It is wrongly imagined that these streams would participate through accelerated velocity in the canyon-cutting activity of the streams coursing down the margins of the uplifted continents, and that the inland areas should retain the record of such erosion in the form of deep valleys far from the coast. But the upper reaches of African rivers (Zambezi, Limpopo, e.g.) wander sluggishly over the level highlands devoid of downcutting power. Vertical uplift without tilt affects stream velocity not at all. The gorge-cutting at the edges of the plateau and that of the submarine canyons starts on the face of the steep slope and extends inland only so fast as the steep slope is notched at its head. Beyond the point of such notching the preuplift regimen of a stream is quite undisturbed.

The reader may consider this continental uplift formulation to be no less speculative than recourse to large variations in sun energy as a reason for glacier-creating refrigeration. There is, how-

Epilogue 129

ever, the difference that in the uplift hypothesis the concern is with observable terrestrial phenomena. Also that apparent examples of the results of such en masse uplift persist in existing relief features. The Colorado Plateau is made up of strata that were deposited at or below sea level and now stand 7,000 feet high. Beyond that is the established fact that the highlands and lowlands of the earth's surface are in isostatic equilibrium, that is, float high or low in accordance with their different densities. On the greatest scale this adjustment is between the rocks of the continental blocks and those of the ocean bottoms. The continents stand high because they are light; the ocean bottoms are floored by denser rock materials. The African plateau, the Venezuelan highlands, and the Colorado Plateau are indicative of deeper-than-average prisms of light continental rocks that were, accordingly, raised to superior heights in Pleistocene times and have sunk back only to preglacial levels in postglacial years. Appropriate to this interpretation is the fact that an ice cap mantled the summit level of the Colorado plateaus during the Pleistocene glacial epoch.

Isostasy requires that below the rigid upper crust of the earth there exist a yielding substratum in which the surface altitude adjustments are made. Moreover, all available evidence indicates that the interior of the earth is hot. Whether this heat is a residue of an original molten earth or is generated by radioactivity or atomic reactions does not directly matter; the high temperatures are indicative of a delicately triggered adjustment and of highly mobile interior conditions.

Accordingly, when consideration is given to the 4,000 mile radius of the earth, it is apparent that a change of one four-thousandth or one two-thousandth (one mile or two miles in length of radius) in surface altitudes is an extremely minor response to potential interior-earth reactions.

Specifically, a relative slight increase in the density of the yielding substratum would cause the continental blocks to float higher and if this increase in density involved a parallel contraction in the material of the ocean bottoms these would sink lower.

By a rise in altitude of the continental blocks of 5,000 to 10,000 feet the first condition for the initiation of glaciers through reduction of the annual average temperatures of the middle latitudes would be adequately met. As has been noted, air temperatures decline at every latitude, 1° F. for every 300 feet increase in altitude. An uplift of 6,000 feet, 20° F. lower temperature, would suffice amply to bring about the refrigeration necessary for the accumulation of continental glaciers on the uplands of the temperate lands. The uplifts, as here conceived, could well be of a pulsating nature and thus account for several generations and dissipations of the ice sheets within the 600,000 to 1,000,000 years that the Pleistocene glacial epoch endured.

Further, it is well established that the wind systems of the present time are the same as they were in the glacial epoch. Thus adequate snowfall is assured in the stormy West Wind belt.

By these means both the cold and precipitation as snow necessary for an ice age are available through terrestrial phenomena without recourse to speculative variations in the radiant energy of the sun. The mysterious submarine canyons are accounted for without assigning extravagant erosive activities to sluggish and chiefly hypothetical subsurface ocean currents.

Appendix I: Vantage Points and Excursions *

MANY of the viewpoints and phenomena listed are described specifically in the text, some at length. They are included here to provide a ready reference for readers who may find a guide useful for planning excursions with definite objectives.

Canandaigua Lake. A fine view and romantic prospect can be seen from the summit of Hatch Hill southeast of Naples. This can be approached by the road, N.Y.S. Route 364, out of Penn Yan but requires a climb to reach the top.

Carpenter Falls. Here is a secluded, little-known, romantic, high, vertical fall over the Tully limestone as the cap rock. On the west side of the Skaneateles Lake valley, it is an unspoiled wooded site for picnics and photographs.

Follow N.Y.S. Route 41A south from Skaneateles village about twelve miles to a dirt road leading east at the North New Hope school. Follow the dirt road 1,500 feet to the crest of the fall. Approached from the south, the site is about six miles from Sempronius on N.Y.S. Route 41A.

Cayuga Lake. An excellent view of length of the lake is to be had from the top of the Library Tower, Cornell University. A notice at the base of the tower gives visiting times. A nice sunset view across the lake may be seen from the crest of the rise on Hillcrest Road leading off from North Triphammer Road south of Asbury Church.

Drumlins. These forms, collectively considered, are impressive as a drumlin "swarm," and the interested observer should undertake a tour that will emphasize this aspect of the phenomenon. N.Y.S. Route 31 west from Jordan and N.Y.S. Route 5 west from Camillus to Lyons (using secondary roads in part) lead through the heart of the drumlin country. Unfortunately there are no high vantage points from which an over-all view can be had. But as the drumlins range from 75 to 200 feet in height above the level lands between them, it is possible by cruising around to locate a high unwooded example from the summit of which there is a prospect of surrounding occurrences. Except along the shore of Lake Ontario, where wave attack has exposed the interior of drumlins, there does not seem to be any example of a drumlin section by a road cut. Even with modern earth-moving equipment, making a cut through a hill 150 feet high and a thousand or more feet wide is a formidable undertaking. So roads go around drumlin ends or even over their tops.

Erratics. The largest erratic of the region known to the author is on the summit of a hill immediately north of the village of Slaterville Springs. Another very large one can be seen on the bed of Fall Creek looking downstream from the bridge at the east end of Forest Home village; a third is the boulder-seat memorial for Professor R. S. Tarr at the southwest corner of McGraw Hall, Cornell University. All three of these are anorthosite masses, a coarsely crystalline rock that crops out over large areas of the central Adirondack Mountains.

Escarpments. Aside from the segmented section of the Portage escarpment, the only expression in the Finger Lakes Region of the presumed preglacial prominence of the cuesta scarps developed by massive limestone beds is that of the Onondaga south of Syracuse on U.S. Route 11.

How significantly such escarpment forms functioned may, however, be appreciated by viewing the Helderberg Mountains scarp south of the Mohawk River valley or the Niagara escarpment at Lewiston.

The forward segment of the Portage escarpment has conspicuous expression when seen, looking southeast, from points at the crest of

Lake clay in cut along Warren Road, altitude 960 feet, showing thick deposit of clay completely free of boulders. (Photograph by J. D. Burfoot, Jr.)

Kame deposit near Freeville, New York. The dry hollow at the base of the tree on the left is the characteristic feature of such deposits. Their gravelly composition is suggested in the partial view of another pit in the lower right-hand corner. (Photograph by O. D. von Engeln.)

Esker at Malloryville, New York. The winding ridge is composed of gravel deposited on the bottom of a subglacial stream. (Photograph by Douglas Rogers.)

Vantage Points and Excursions

the hill on N.Y.S. Route 96 going north toward Trumansburg from Ithaca. It is also prominent south of N.Y.S. Route 90 between Dryden and Messengerville.

Hanging valleys. Go out toward Mecklenburg on N.Y.S. Route 79 from foot of State Street, Ithaca, to the end of the big curve in the road up the hill. From there the hanging valleys of Fall Creek, Cascadilla Creek, and, best, Sixmile Creek can be seen in relation to the Cayuga–Inlet Valley trough.

On N.Y.S. Route 281 at Tully village, looking west across the Onondaga Valley, the hanging valley of Vesper Creek is conspicuous.

At the junction of N.Y.S. Route 13 and N.Y.S. Route 224 near Cayuta village, a hanging valley on the west side of the Cayuta trough is prominent.

High points. An unnamed forested hill, altitude 2,133 feet, southeast of the village of Virgil on N.Y.S. Route 90 is the highest summit in the Finger Lakes "East" area.

However, Connecticut Hill, 2,095 feet, is usually credited with this distinction. From Ithaca follow N.Y.S. Route 79 toward Mecklenburg to the junction with Route 327, then South on 327 to Trumbull Corner, and then straight west on the road from Trumbull Corner to the summit. The view from the top is somewhat restricted because of the growth of tall brush.

The summits in and south of the Finger Lakes "West" are considerably higher than these. Gannett Hill west of Canandaigua Lake above the village of Bristol Springs rises to 2,256 feet, the highest summit in these parts. Call Hill, about six miles south of Hornell on the road up Purdy Creek to Andover, is 2,401 feet high and not wooded.

Interglacial and postglacial gorges. For viewpoints from which both the postglacial and interglacial gorges of Buttermilk Creek can be seen *in toto:* go out Elm Street Extension, west from the west end of State Street, Ithaca. From several points at an altitude above 1,000 feet the length of both gorges may be seen, the interglacial gorge on the north (best seen when the trees are bare of leaves). This appears to be the one instance where the gorges are parallel and not intertwined and where also the interglacial gorge was left nearly unfilled by deposits of the second glacierization. This is convincing evidence of two glacial advances with an interglacial period intervening.

Kettle ponds. Picturesque water-filled kettles occur less than one mile north of the village of Ellis (Hollow), on the north side of the Ringwood Road.

An exceptionally large such pond with pleasant prospects along its shores is Dryden Lake. This can be circled by following N.Y.S. Route 38 south out of Dryden village.

N.Y.S. Route 80 west from Tully village skirts a number of smaller kettles on the back, south side, of the Tully moraine. Farther south along N.Y.S. Route 281 are the larger Little York outwash plain kettle lakes.

Peneplain uplands. In general this designation is given to the country beyond the summit of that segment of the Portage escarpment south of Catatonk Creek between Van Etten and Candor. Roads to the south from N.Y.S. Route 223 will afford characteristic, though considerably dissected, prospects over this aspect of Finger Lakes Region topography.

A less-dissected tract occurs south of Andover (N.Y.S. Route 17); take the first road to the south, east of Andover.

Rock hills modeled by glacial erosion. A side road to the east branching off from N.Y.S. Route 96 at Interlaken affords views of the glacially modeled upland topography on the far, east side of Cayuga Lake. The rounded, boat-bottom form of these hills is characteristic of the ice erosion of rock structure present in the Finger Lakes Region.

Seneca Lake. A good view of the lake is to be had from the cemetery on the north side of Watkins Glen. Descend from the lunch pavilion at the upper end of the glen, keeping above the gorge walls. There is no path. The cemetery occupies the terraces of the hanging deltas of Glen Creek.

Skaneateles Lake. Approach from the south over N.Y.S. Route 41 to a point just beyond the southern end of the lake for the view referred to as the Switzerland of America. The road to the west from Scott (on Route 41) leads down to a swampy delta at the head of the lake. Here there are four or five enormous hemlock trees, the only known survivors of the primeval forest of the Finger Lakes Region.

Taughannock Falls. The best over-all view is from the lookout site off from the road on the north side of the gorge. The most impressive view within the gorge is looking up just below the falls. Be on guard against falling rocks.

Vantage Points and Excursions

Visit the falls after a period of heavy rain. The ice dome below the falls after a long period of cold late in the winter is very spectacular.

Through valleys. This feature has its most striking expression in the Cayuga Inlet valley south of Ithaca, N.Y.S. Routes 13 and 34. The similar development south of Seneca Lake, Watkins Glen to Elmira, is of comparable size, but the erosional aspects are not so prominent.

Although not related to larger drainage lines, Labrador Hollow, south of Apulia, N.Y.S. Route 91, and Texas Hollow, south of Bennettsburg, at the junction of the road with N.Y.S. Route 227, are convincing, even spectacular, demonstrations of the glacier's effectiveness in channeling erosion.

Tinker Falls. This secluded, romantic spot, not generally known, is on N.Y.S. Route 91 south from the village of Apulia, in Labrador Hollow, a short distance up a gorge extending east from the road opposite Labrador Pond on the floor of the valley. The Tully limestone overhang of the fall is a notable feature.

Truncated spurs and slopes oversteepened by glacial erosion. Impressive instances are seen from (1) U.S. Route 11 looking across the valley near Tully at Toppin Mountain; (2) east of Elmira Heights from N.Y.S. Route 328 south of Horseheads; (3) the east side of the south end of Canandaigua Lake, from N.Y.S. Route 245 out of Naples, New York.

· *Appendix II: Maps* ·

JUST preceding the References is an index map of the topographic maps of the Finger Lakes Region, published by the United States Geological Survey.

When ordering maps, it should be requested that they be sent rolled. Each map costs 30¢, but if at the 30¢ price an order amounts to $10.00 or more a 20 per cent discount is allowed; thus a $10.00 order would cost only $8.00. Order by the name of the map and state, e.g., Tully N.Y. Order from U.S. Geological Survey, Washington 25, D.C. Remittance by check or money order, must accompany the order. Make remittances payable to United States Geological Survey.

With one exception these maps show relief only by contour lines. The exception is the Ithaca-West sheet, which is also furnished (same price) in shaded relief in brown and shows the topography in perspective.

A single over-all map, 100-foot contour interval, of the Finger Lakes Region is called Elmira sheet, United States Series of Topographic maps, Scale 1: 250,000. This sells for 50¢.

The Army Map Service, Washington, D.C. supplies this map under the designation "Elmira, N K 18–4" as a plastic relief map. This is in effect a relief model of the area with the vertical exaggerated 3X. Its price is $3.50. The best bird's-eye view representation of the region with all the salient features of the relief visualized in a glance, it is an assured conversation piece where displayed on the wall of a study, office, or lobby.

Glossary

MOST of these terms are explained where they are first used in the book but are included here for convenient reference and to provide formal definitions.

Ablation moraine. Rock debris in transport on the surface of a glacier, or such debris let down to the ground from the surface or from the interior of the ice when the glacier finally melts.

Aggraded. If a stream at any part of the course has more or coarser sediment than can be moved farther by the current it aggrades its bed, that is, deposits the excess of load on the bottom.

Anticline. Layered rocks bent upward in arches of small to mountainous dimensions.

Aphelion. The point in its orbit where the earth is farthest from the sun.

Barbed tributary. If the course direction of a tributary where it joins the main stream is opposite to that of the main stream it is called a barbed tributary by analogy with the barb of a fish hook. The phenomenon is evidence of a reversal of flow direction in the valley of the main stream.

Bedrock. Solid rock which, either exposed to the air or under a soil cover of varying depth, is everywhere present below the surface of the earth.

Boulder clay. Deposited ground moraine, so named because characteristically it is a mass of rock flour studded with boulders.

Caprock. Resistant, horizontal or nearly horizontal, rock layer at the crest of an escarpment or the brink of a waterfall. The escarpment or waterfall is maintained as such because the top surface of the resistant bed persists against weathering and erosion.

Consequent stream. Stream that owes its course direction and characteristics to original land slopes; that is, it originates in consequence of the introduction of such new slopes.

Continental glacier. An ice mass of continental dimensions of such great thickness that neither its form nor its spread is conditioned by the relief of the land surface over which it moves.

Contour interval. The vertical distance indicated by any two successive contour lines on a topographic map. The gentler the slope of the land, the farther apart successive contour lines will be on the map. See *Contour line.*

Contour line. A line on a topographic map connecting all points of equal elevation.

Crevasse. With reference to glaciers an approximately vertical open crack extending down from the surface of the ice into the interior of the glacier. The maximal depth of crevasses is about 200 feet.

Crop out. Verb form for *Outcrop* (which see).

Crystalline rocks. Rocks composed of mineral crystals. Such rocks originate either from cooling of molten rock substance or from conversion of noncrystalline rocks to the crystalline condition by application of heat or pressure or both.

Cuesta. A hill with one steep (escarpment) slope, the other gentle (dip slope). A land form developed by stream dissection of gently inclined layered rocks composed of thick units differing greatly in degree of resistance to such attack.

Cut bank. While ingrown meanders (which see) are being developed, the side of the valley against which the curved flow is directed experiences lateral undercut erosion. The materials above, deprived of support, fall into the stream and are carried away. Thus a steep wall, a cut bank, is maintained at those places in the stream course.

Glossary

Delta. An accumulation of sediment deposited by a stream on entering a body of still water, a lake, or the sea. Since the current ceases at the stream mouth, its load of rock debris is dropped at the shore and in time may build up a wide, thick land mass.

Dendritic drainage. A pattern of drainage that simulates the organization of a tree, such as an oak, into trunk, branches, and twigs.

Distributary valley. The volume of water or ice from a tributary valley augments the main stream. Distributary valleys partition the flow of the main current into two or more lesser divergent currents.

Drumlin. A hill composed of ground moraine, in form resembling an overturned tea spoon with the blunt end opposed to the direction of ice advance.

End moraine. Ridge of glacially eroded rock debris deposited at the end of a glacier, specifically at the end of a projecting lobe of the ice.

Englacial moraine. Rock debris from glacial erosion in transport in the body of the ice.

Epoch. In geology an interval of time shorter than a period, longer than an age, in the sequence era, period, epoch, age.

Erosion. In geology the mechanical wearing away of rocks by the grinding and tearing of moving agencies—rivers, glaciers, wind.

Erratic. Any boulder (but usually used with reference to a large one) deposited by a glacier at a point far removed from the outcrop of the parent rock from which it was derived—in the Finger Lakes Region, all crystalline boulders.

Escarpment. As used in this book, a cliff whose crest is determined by the terminal outcrop of a massive, resistant stratum, commonly limestone, or to a close-spaced vertical sequence of durable thin beds, usually sandstone. Escarpments are etched into relief by weathering and stream erosion.

Esker. Gravel deposit forming a linear ridge, made on the bottom of a subglacial stream flowing under ice stagnated at the front of a glacier.

Faceted. Used to designate plane surfaces developed on pebbles used by a glacier as a grinding tool.

Fault. In geology a break, straight or curved, in consolidated or unconsolidated rock materials or in ice, with movement of translation in the surface of the break.

Fiord. Narrow valley with high, steep sides extending a long distance inland from the sea with its bottom far below sea level, hence an arm of the ocean. Product of glacial erosion.

Formation. A rock unit that has recognizable characteristics so that it may be traced or identified when encountered at different localities.

Geomorphology. The word translates literally to science of the form of the earth. Used to designate interpretations of the origin and development of the relief features of the land.

Glacial epoch. See *Pleistocene epoch.*

Glacial lake clay. The rock flour (which see) of glacial grinding present as a deposit on the bottom of a proglacial lake.

Glacier. Ice mass formed from accumulated snow which flows under the pressure of its own weight.

Glacier milk. Milk-white stream flow, melt water from glaciers, due to a heavy load of suspended sediments composed of the rock flour made by the glacier's grinding rocks into a white powder.

Glacierization and glaciation. Glacierization means the invasion and occupation of a region by a glacier or glaciers; glaciation refers to the effects, changes, brought about by glacierization.

Gneiss. A granite or sandstone rock that has been converted by heat and pressure to a banded crystalline rock.

Granite. A coarse-grained crystalline rock which has a large amount of quartz in its composition. In popular usage many varieties of coarse-grained crystalline rocks are called granite.

Ground mass. The fine material present between fragments or crystals of rocks large enough to be identified; for example, the rock flour between boulders in boulder clay.

Ground moraine. Debris of glacial erosion in transport at the base of the glacier, or deposited from the base of the ice.

Hanging valley. A tributary valley whose bedrock floor at the site of the junction with a main valley is appreciably above, commonly high above, the floor of the main valley. Ordinarily, tributary streams join main streams at equal levels.

Hardpan. A tenacious and refractory subsoil layer of various orgins; in the Finger Lakes Region ground moraine.

Headwater erosion. The disintegrating action of a stream on the rocks of its source area. By this process a stream valley is lengthened at its head.

Hinge line. In connection with uplift of the land it has been found

Glossary

that in certain instances the uplift proceeds as when a trap door hinged on one side in a floor is raised. The surface area of the door is tilted, but beyond the hinge line the level of the floor is not affected.

Hooked tributary. Another term for *Barbed tributary* (which see).

Ice Age. See *Pleistocene epoch.*

Incised meander. See *Ingrown meander.*

Ingrown meander. A slow-moving stream regularly pursues a winding course, that is, meanders from one side of its valley floor to the other. If, then, the stream gradient is suddenly increased, the stream will cut downward as well as laterally and the meander pattern of flow will be engraved below the former level of the valley. Meanders are then ingrown.

Isostasy. The principle that the major relief features of the earth's surface float high or low in a yielding substratum according to their density, other factors being equal.

Joint plane. A minute fissure through rock masses bounded by plane surfaces. Unlike faults there is no displacement by movement in the plane of a joint.

Kame. Extremely hummocky, commonly massive, deposits of sand and gravel accumulated from water transport at the front of a glacier; typically deposits at the outlet of an esker (which see) stream.

Kettle. A closed depression in a deposit of moraine or kame resulting either from the unequal dumping of the debris or because an ice block, buried in the moraine or kame, melted out after deposition had ceased.

Lacustrine deposits. Sediments deposited on a lake bottom.

Limestone. Sedimentary rock composed of calcium carbonate.

Local base level. The level below which a stream cannot deepen its bed at a particular locality. This may be governed by the main stream into which it flows or the level of a lake into which it empties.

Melt water. Water derived from the melting of ice in the terminal area of a glacier.

Migration down dip. The principle that the courses of streams flowing over rock beds inclined from the horizontal are progressively shifted laterally in the direction of the dip except in cases where the flow is exactly parallel to the dip direction.

Moraine. Rock debris collected and transported by glaciers either while still being carried by the ice or after being deposited by and from the ice. Deposited moraine may appear as sheets, heaps, or ridges.

Outcrop. Appearance of the solid bedrock at the surface, especially if the exposure is only of limited extent. Commonly referred to as a rock ledge.

Outwash gravel. Gravel deposited from melt-water streams immediately beyond a glacier front. Streams that flow off a glacier have steep gradients. Those which emerge from under the ice flow under pressure. In either case their velocity is checked when they leave the ice. Since such streams are heavily charged with sediment, a significant check in velocity compels deposit of part of the load.

Outwash plain. A broad expanse of gravel and sand deposited by streams outflowing from the front of a glacier.

Peneplain. A wide expanse of country of low relief elevated only slightly above sea level. This nearly plane surface extends indifferently across diverse kinds of rock and geologic structures, all of which have been reduced to a common level by weathering and stream erosion. The low relief may be preserved after uplift as the summit level of a plateau.

Perihelion. The point in its orbit where the earth is nearest the sun.

Pitted outwash plain. When blocks of ice buried under the gravels of an outwash plain melt after deposition has ceased, the overlying gravel beds slump down and so create pits in the surface of the plain.

Pleistocene epoch. The geologic time interval of the continental glacierization preceding present time. In this book, synonymous with Ice Age and glacial epoch. Computed to have terminated in the Finger Lakes Region 9,000–10,000 years ago and to have extended through 1,000,000 years.

Plucking. In glacial geology the process of erosion operating when the bottom ice wedges, pries, or pulls off substantial fragments of bed rock.

Plunge pool. Pool excavated in bedrock at the foot of a vertical waterfall by the impact of the descending water and swirling currents in the pool, using sediments as grinding tools.

Pothole. An approximately circular depression in the rock floor of a stream bed increasing in diameter from the surface downward. An initial depression in the stream bed induces a rotary motion in part

Glossary

of the current, and the swirl, using sand, pebbles, and boulders as tools, grinds the hollow deeper and wider.

Proglacial. In front of a glacier terminus—used with reference to lakes ponded by the ice dam.

Quartzite. A pure sandstone, all silica, which has been converted by heat and pressure to dense rock which shows little or no trace of the original sand grains.

Recessional moraine loop. Ridge of glacial debris, extending across a valley, convex in the direction of ice motion, and marking the position and outline of the ice front during a halt in the final melting away of the ice.

Regional dip. The inclination of beds of sandstone, shale, and limestone away from the horizontal. Regional dip is in part the initial inclination of the sediments as deposited on a sloping sea bottom, in part acquired during their uplift above sea level.

Rock-defended terrace. When a meandering stream is maintaining a cut bank (which see) by cutting laterally into unconsolidated materials, it leaves a gently sloping terrace on the other side of the stream. Meanders migrate downstream. The next meander has the opposite bend, hence operates to develop the cut bank in the terrace area. But if a rock spur intervenes, the meander is prevented from cutting away all the terrace area. The part preserved is then called a rock-defended terrace.

Rock flour. Rock powder of flourlike fineness produced by glaciers grinding off fresh rock from the surfaces over which they pass.

Sandstone. Consolidated rock in a bed, composed of grains of sand cemented together.

Sediment. In geology rock particles, pebbles, sand, clay in transport, primarily by streams, or deposited in layers on the bottoms of lakes and the sea. Such deposit occurs when the current no longer suffices for their transport farther.

Sedimentary rock. Rock made up of sedimentary beds in some degree consolidated by compaction or cementation.

Seismic sounding. Through minor earthquakes artificially produced by underground explosions, soundings can be made of the depth of loose fill over a valley floor or of the depth below the surface of distinctive rock beds. Greek *seismos* is earthquake.

Shale. Consolidated rock, in beds, composed of clay particles made coherent by compaction due to the pressure of overlying formations. Very little resistant to disintegration from weathering or destruction by erosion.

Solution pit. An irregular hollow in the top surface of a limestone that is a stream bed. The purest parts of the limestone are most rapidly dissolved by the flowing water, thus depressions develop over those areas.

Stratigraphic sequence. The order in which sedimentary beds are arranged according to the times of their successive deposit, the younger overlying the older.

Stratigraphy. Geologic description and correlation of the stratified rocks.

Stream capture. The diversion of a stream from an earlier course to become a tributary to a stream flowing at a lower level when the headwater erosion of the latter intercepts the course of the other stream.

Striated. Used to describe the linear scratches on bedrock and on boulders made by the grinding action of the bottom ice of an actively moving glacier.

Subaerial. Used to indicate that the geological action referred to takes place in contact with the atmosphere.

Sublacustrine moraine. A moraine ridge built up under water on a lake bottom.

Synclinal axis. Where rock beds are bent into arches and troughs either of small scale or of mountainous dimensions; the downbends of the folds are called synclines. (The upbends are anticlines.) A line parallel to the bottom of the synclinal trough is its axis.

Terminal moraine. Glacially derived rock debris accumulated by dumping at the end of a glacier during a period when the ice front was maintained at the same line—in careful usage refers only to deposits made at the line of the farthest advance of the ice.

Through valley. A valley with its divide between drainage flowing in opposite directions on the valley floor instead of on a ridge. In the Finger Lakes Region north-south valleys whose preglacial ridge-divides have been reamed out by glacial erosion.

Till. Deposited ground moraine.

Topographic survey. A survey for map-making that, besides locating all features of the landscape, also gets data for representation of the relief, hills, plains, valleys, of the surface—its topography.

Glossary

Truncated spur. Long spurs typically extend in alternation from the sides of valleys of gentle gradient created by stream erosion. Massive ice currents moving through such valleys cut off the ends, that is, truncate the spurs.

Unconsolidated material. The soil and, in places, great thicknesses of material below it lack coherence, are unconsolidated. At depth, however, solid bed rock is always encountered.

Valley fill. A general term used to refer collectively to the several types of glacial debris, till, moraine, outwash, etc., deposited over the bedrock of a valley bottom.

Valley train. When the deposit of outwash gravel from a glacial stream is confined to the floor of a relatively narrow valley, the long-extended accumulation is called a valley train.

Weathering. The collective term for the processes of decomposition and disintegration (solution, hydration, oxidation, chemical reactions, heat and cold, frost) that bring about the decay and mechanical destruction of consolidated rocks where these are exposed to the air and percolating water.

Index map for the United States Geological Survey maps.

References

REGRETTABLY nearly all government reports dealing with the landscape features of the Finger Lakes Region are technical in content, and articles in popular journals are extremely superficial. Unfortunately also, such books as do have chapters or sections with serious descriptions of the region are mostly out of print.

Nevertheless it seems desirable to append this list of references to indicate the nature and scope of the available literature on the subject. The out-of-print books will commonly be available in public and college libraries.

Tarr, R. S. *The Physical Geography of New York State.* New York: Macmillan Company, 1902.

Although written for the layman, this book contains references to all the technical literature on Finger Lakes landscape features up to the time of its publication. It was at about this time that the long controversy, glacial-erosion origin versus stream-erosion origin of the Finger Lakes basins, was finally resolved, in favor of glacial erosion. Later literature in general recognizes the efficacy of the ice attack.

Fenneman, N. M. *Physiography of Eastern United States.* New York: McGraw-Hill Book Company, 1938.

This is a technical account. The Finger Lakes Region is given only brief consideration, but references to technical papers are brought up to the date of publication. Diagram maps show the lower level, later stages of the ice-dammed proglacial lakes.

Bowman, I. *Forest Physiography*. New York: John Wiley and Sons, 1911.

This book relates the land forms of the United States to the natural forest and vegetative growth of the country. The last chapter "Lowland of Central New York" is a concise description of the Finger Lakes Region.

Atwood, W. W. *The Physiographic Provinces of North America*. Boston: Ginn and Company, 1940.

The most recent, generalized, popular account of the relief forms of the continent and their development. Only a few pages are devoted to the Finger Lakes Region.

The listings above are all the books that include generalized popular or semipopular accounts of the Finger Lakes Region. Aside from uninformative pieces in magazines, the rest of the literature, though voluminous, is almost exclusively technical. There are, however, certain of these items that should be listed because they constitute source material for anyone doing research in the natural history of the area.

Birge, E. A., and Juday, C. *A Limnological Study of the Finger Lakes of New York*. (Document No. 791, issued October 27, 1914; from Bulletin of the Bureau [U.S.] of Fisheries, Vol. XXXII, 1912.) Washington, D.C.: Government Printing Office. 85 pages and maps.

This study is a comprehensive, analytical description of the topography and hydrography of the lakes—size, volume, depths, seasonal temperatures, dissolved gases, water circulation, etc.—and includes individual contour maps of the lakes on the scale 1 inch to the mile. It is the source paper for all numerical data in this book relating to the lakes.

Williams, H. S. Tarr, R. S., and Kindle, E. M. "Watkins Glen–Catatonk Folio" (Folio 169), *Geologic Atlas of the United States*. Washington, D.C.: U.S. Geological Survey, 1909.

This is the only basic, field-research paper on the geology and geomorphology of a representative portion of the Finger Lakes Region. It covers the southern parts of the Cayuga Lake and Seneca Lake basins and adjoining areas. The study is supplemented by colored maps on the scale of ½ inch = 1 mile, showing the topography, geology, and distribution of glacial deposits.

This is a technical treatise intended primarily for professional readers. Unfortunately the organization is involved, and, as no index is provided, it is a difficult paper to consult.

Carney, Frank. *The Pleistocene Geology of the Moravia Quadrangle, New York.* (Reprinted as a separate publication from the Bulletin of the Scientific Laboratories of Denison University, Vol. XIV, pp. 335–442.) Granville, Ohio: Denison University, 1909.

This, like the Watkins Glen–Catatonk Folio 169, is a technical field-research paper describing the glacial phenomena of the area of the Finger Lakes Region south of Owasco Lake. A black-and-white map of the glacial features is appended.

von Engeln, O. D. "The Finger Lakes Region," XVI International Geological Congress, Guidebook 4, Excursion A-4, pp. 39–69. Washington, D.C.: Superintendent of Documents, 1932.

This contains a concise resumé of the geology and geomorphology of the region, followed by itineraries of excursions to significant stations in the area. Some of the interpretations made at that time differ from those in this book.

The appended bibliography includes most of the pertinent technical literature on all aspects of the geology of the region.

von Engeln, O. D. *The Tully Glacial series.* (New York State Museum Bulletin 227, 228, pp. 39–62.) Albany, N.Y.: 1921.

This account is phrased to acquaint the lay reader with the imposing sequence of glacial phenomena displayed in the Onondaga–Tully valley. It includes a map and photographs.

Watson, T. L., *Some Higher Levels in the Postglacial Development of the Finger Lakes of New York State.* (New York State Museum, in Fifty-first Annual Report, pp. 55–117). Albany, N.Y.: 1898.

The first systematic study of the succession of proglacial lake levels of the Finger Lakes "East," this is still the basic reference.

Wold, J. S., "Interglacial Consequent Valleys of Central New York," *American Journal of Science,* CCXL (1942), 617–626.

This article gives a technical but very lucid account of this significant phenomenon.

· Index ·

Page numbers in italics refer to illustrations.
All place names refer to New York State unless otherwise indicated.

Adirondacks, 2
Alpine, 114
Appalachian Mountains, 3-4, 13, 33
Apulia, 34, 52, *53*, 65
Bald Mountain, 91
Batavia, 98
Beebe Lake, 79
Bennettsburg, 64, 110
Big Flats, 84
Big Stream, 59
Blodgett Mills, 42, 67
Boulder clay, 103
Boulders, flatiron, 103
Brooktondale, 64
Buttermilk Creek, 109
Buttermilk Falls, 77
Buttermilk Gorge, *68-69*, 77
Canadice Lake, 21, 27, 39, 40
Canandaigua Lake, 21, 27, 39, 40, 131
Canasawacta Creek, 36, 37
Candor-Van Etten valley, 24, 43
  see also Catatonk Creek
Canisteo River valley, 115
Cascadilla Creek, 26, 58, 66, *68-69*, 77, 94, 95
Catatonk Creek, 22, 43
  see also Candor-Van Etten valley
Catherine Creek, 60
Catlin Mill Creek, 59
Cayuga Inlet valley, 24, 25, 58, 63, 94, 108, 110
Cayuga Lake, 131-132
Cayuga Lake valley, 57-58, 89, 109
  preglacial drainage of, 16, 20
  rock cliffs of, 92
Cayuga "River," 20
Cayuta Creek, 22, 115
Cayuta Lake, outlet gorges of, 61, 62, 80-83, *82*, *83*, *84*, *85*
Cayutaville, 62
Cazenovia, 18
Centerlisle esker, 119
Chemung River, 80, 83-85, 96
Chenango River, 35, 36, 37, 41-42
Clark Reservation, 98-99, *98*
Cohocton River valley, 115
Conesus Lake, 27, 39-40
Connecticut Hill, 34, *34*
Consequent streams, 71
Cornell University campus, 94
Corson, Hiram, 77
Cortland, 41, 43-44, 66, 116
Coy Glen, *84*, 86-87
Crooked Lake, 39
Cuesta, 12, *12*, 15, 16
Danby, 61
Davis, W. M., 62
Delaware drainage, 4
Deltas:
  at Brooktondale, 91-92
  at Taughannock, *68-69*, 74
  hanging, *84*, 86
Deposits by Ice Age glacier, 10, *101*, *132*
Divides:
  between north-flowing and south-flowing drainage, 5, 14
  post-peneplain uplift, 18, 20
  preglacial, 41-42, 59, 64

153

Doolittle Creek, 25-26
Drainage:
  postglacial Cayuga, *24*
  preglacial Cayuga, *25*
  south-flowing, 36-37
Drumlins, *101*, 104-105, 132
Dryden, 27, 41, 66
Dryden Creek, 44, 117
Dudley Creek, 119
Ellis Hollow, 67, 91
Ellis Hollow Road, 66, 94
Elmira, 84, 114
Enfield Glen, *69*, 77-78, 80
Erosion, glacial, 8-9, 46-48
Erratics, 50, 52, 106, *116*, 132
Escarpments, 132-133
  see also Helderberg, Niagara, Onondaga, Portage, and Tully escarpments
Eskers, 117-118, *133*
Factory Brook, 45
Fairchild, H. L., 48
Fall Creek, 22-23, 26, 41, 42, 43-44, 45, *68-69*, 73-74, 79, 87, 100, 107
Fall Creek valley, 23, 26-27, 58, 94-95, 99, 107, 109, 117, 118
Finger Lakes:
  depths of, 53
  levels of, 88
  rock basins of, 52-55
Finger Lakes "East," 27, *29*, 39, 40, 43, 55, 60
Finger Lakes Region:
  average annual temperature of, 123
  beginning of geologic history of, 2
  cuesta sculpture of, 15
  divide between north-flowing and south-flowing drainage, 14
  gorges and submarine canyons, 126-127
  invasion by continental glacier, 49-51
  postpeneplain divide in, 20
  postpeneplain drainage development of, 5-6
  preglacial drainage in, 59
  preglacial drainage level of, 58
  role of escarpments in, 16
  scenery of, 119-120
  sequence of landform evolution of, 27-31
  thickness of ice cover in, 50
  thickness of layered rocks in, 3, 13, 27
  topographic maps of, 94
  topography from ice erosion and glacial deposits, 9-10
  uniqueness of, 2, 3, 38-39, 55, 68
  uplifts of, 3, 5, 13, 14
Finger Lakes "West," 27, *28*, *36*, 39, 40, 42, 43, 54-55, 101
"Flat Rock," 22, 100
Flood of 1935, 75
Forest Home, 22, *68-69*, 79, 95, 99, 100, 106
Freeville, 116-117
Gill, A. C., 48
Glacial deposits, nature of, 102-103
Glacial erosion, 51-53, *52*, 54-55
Glacial lake clay, 111, *132*
Glacial outwash, 111-112
Glacial period, *see* Ice Age
Glacier, continental:
  interior currents of, 50-51
  map of, *frontispiece*
  melting of, 5, 89, *116*
  milk, 104
  motion of, *4*, 9-10, 48-49, 56, 63
  thickness of, 89, *116*
Glacierization, multiple, 70-71
Glen Creek, *20*, 59, 87
Gorges:
  interglacial, *68-69*, 78-80, 133-134
  outlet, 60-61, 80-83, *82*
  postglacial, *20*, 68, *68-69*, 71-72, 73-74, 75, 78, 133-134
Groton, 109
Halsey Valley, 42, 65
Hanging valleys, 57, 60, 133
Harford, 41, 42, 45
Harz Mountains, 38
Havana Glen, 59
Hector Creek, 109
Hector Falls, 59
Helderberg escarpment, 17, 22
Hemlock Lake, 21, 27, 39, 40
Hendershot Gulf, *82*, 83, *84*, 90
High-level lakes, 60, *85*, 86-100, *100*
High points, 133
Hoddy-doddies, 104
Homer, 18, 65, 114
Honeoye Lake, 27, 39-40
Hooked tributary, 21, *21*, 36-37, 40, 42-43, 59
Horseheads, 64, 84-85, 96, 97, 114
Ice, thickness of, 67, 70
Ice Age, 6-7, 9, 69-71, 121-126
  map of, *frontispiece*
Ice Age glacier:
  thickness of, 89, *116*
  temperature at border of, 8
Ithaca, 58
Ithaca Falls, *68-69*, 73

Index

Jamesville, 52, 98, 99
Johnson Hollow, 60
Kames, 115-116, 117, *132*
Kettles, 107, 113, 115, 117, 134
Keuka Lake, 21, 27, 36, 39-40, 41, 55
Labrador Hollow, *53*, 65
Lake Brookton, 93, 94, 95
Lake Hall, 98, 100
Lake Iroquois, 98
Lake Ithaca, 93, 95, 96, 97
Lake Newberry, 97, 100
Lake Onondaga, 98-99
Lake overflow channels, 89-90, 91, 95, 96-97, 98-99, 111
Lake Slaterville, 91
Lake Watkins, 96, 97
Lake West Danby, 90, 93, 94
Lisle, 119
Lockwood, 22, 42-43
"Lost Gorge," 81, *82*, *83*, *84*, 90
Lowland embayment of Central New York, 18-20
Ludlowville, 97
Malloryville esker, 118
Mandana, 18
Maps, 137-138
  index map of topographic maps, *148-149*
Marathon, 66
Mecklenburg, 58, 61
Mecklenburg Road, 64
Messengerville, 27, 41, 45, 67
Michigan Creek, 62
Michigan Hollow, 61, 90
Millport, 26, 42, 60, 64
Moraine:
  ablation, 106, 112
  building of, 108, *117*
  end, 106
  ground, *101*, 103
  interglacial, 85
  of the "Second Glacial Epoch," 106, *117*
  recessional, 106
  terminal, 10
  "Valley Heads," 106
Mott Road, 117
Mount Pleasant, 58
Nedrow, 65
Newfield-Pony Hollow valley, 26
Niagara escarpment, 17, 22
Niagara Falls, 73, 74, 99
Niagara River, 89-90, 111
North Cohocton, 114
North Lansing, 96

North Spencer, 63, 90, 108, 114, *117*
Norwich, 36
Odessa, 64
Old Salt Road, 118
Onondaga escarpment, 17-18, 22, 99, 105
Onondaga Lake, 98-99
Onondaga Valley, 39, 65, 90, 99, 108, 110
Otisco Lake, 39, 44
Otisco Lake valley, 114
Otselic River, 35, 42
Outwash plains, 113
Overflow channels, *see* Lake overflow channels
Owasco Lake, 18, 39, 44, 97
Owasco Lake preglacial river, 26-27
Owasco Lake valley, 109, 117
Owego, 66
Owego Creek, 25, 27, 45, 66, 91, 119
Owego Creek valley, 115, 117
Peneplain, 4, 13, *14*, *21*
Peneplain uplands, *21*, 32-37, *34*, *117*, 134
Pine Valley, 42, 114
Pleasant Grove Brook, 95
Pleistocene glacial epoch, *see* Ice Age
Plunge pool, *68-69*, 75
Pony Hollow, 110, 114
Portage escarpment, 18-19, 22-23, 24-25, 26-27, 34, 40, 54, 58, 60, 63, 100
Potholes, *68-69*, 76, 77
Proglacial lakes, 60, *85*, 93-94, 96, *100*
References, 149-151
Regional dip, 12
Reynoldsville, 110
Richford, 27, 41, 44-45, 66, 119
Ringwood Road, 23
Roads:
  Ellis Hollow, 66, 94
  Mott, 117
  N.Y. State routes:
    Route 13, 23, 65, 81, 86, 95, 109, 110, 116
    Route 14, 42, 64, 114
    Route 17, 114
    Route 17E, 84
    Route 21, 114
    Route 34, 42, 96, 97
    Route 34-96, 63, 90, 108
    Route 38, 27, 41, 42, 66, 109, 116
    Route 41, 90
    Route 63, 114
    Route 79, 58, 62, 64, 66, 94, 119
    Route 80, 52
    Route 91, 34, 65
    Route 96, 42, 61, 65
    Route 96B, 64

Roads (cont.)
　Route 173, 52, 98
　Route 223, 24
　Route 224, 81, 114
　Route 227, 64, 110
　Route 228, 64
　Route 322, 22
　Route 330, 64
　Route 392, 95, 99, 100
Old Salt, 118
Ringwood, 23
U.S. Route 11, 27, 41, 42, 65, 66, 90, 108, 114
Warren, 95
Salmon Creek, 20-21, *21*, 40-41, 42
Salmon Creek valley, 59, 97
Salt layers, origin of, 3
Scott, 45
Seneca Falls, 18
Seneca Lake, 134
Seneca Lake preglacial river, 26
Seneca Lake valley, 57-58, 59-60, 64, 89, 96
Seneca River, 18
Sixmile Creek, 26, 91
Sixmile Creek interglacial gorge, 78
Sixmile Creek valley, 52, 58, 64, 91, 93, 95
Skaneateles Lake, 18, 39, 44, 45, 97, 134
　high-level overflow of, 90
　preglacial river, 26-27
Skaneateles Lake valley, 114
Slaterville Lake, 91, 93
Slaterville Springs, 66, 91, 106
Solution pits, 74
South Hill, Ithaca, 50, 93, 94
Spencer, 65
Striations, glacial, *37*, 47
Susquehanna drainage, 4, 34-35
Syracuse, 99
Tarr, R. S., 48, 106, 132
Taughannock Falls, *68-69*, 134-135
Taughannock Gorge, *68-69*, 74-75
Temperature at border of Ice Age glacier, 5, 8
Terraces, rock-defended, 100-101
Texas Hollow, 64-65, 67, 110
"The Steps," *84*, 86
Thickness of Ice Age glacier, 7, 46-47, 67, 70
Thousand Islands, 51
Through valleys, *52*, *53*, 62-68, 135
Till, 103
Time:
　for erosion of Fall Creek valley, 44

geologic computation of, 2
interglacial, 72
of halt for end-moraine deposit, 107
of Ice Age, 9, 70, 100
of proglacial lakes, 92
of waterfall at Clark Reservation, 99
postglacial in Finger Lakes Region, 12, 70, 92
when geologic history of Finger Lakes Region began, 2
Tinker Falls, 135
Tioughnioga River, 18, 27, 34, 41, 43, 45, 90, 99, 113
Tioughnioga River valley, 65, 66, 114, 115, 116
Topography of Central New York, 45
Treman, Robert H., 77
Trent River, Ontario, 17
Triphammer Falls, 79
Trumansburg Creek valley, 109
Truncated spurs, 65, 135
Truxton, 18, 34, 65
Tully, 65, 90, 108, 109, 110, 114
Tully-Cortland outwash plain, 113
Tully escarpment, 18, *53*
Tully Lakes, 113
Turkey Hill, 58, 94
"Valley Heads Moraine," 106, *117*
Valley trains, 114-115
Varna, 23, 99, 107
Varna moraine, 95, 109
Warren Road, 95
Waterfalls:
　Buttermilk, 77
　Cavern Cascade, *68-69*, 76
　"fossil waterfall" at Clark Reservation, 98-99, *98*
　Hector, 59
　Ithaca, *68-69*, 73
　Niagara, 73, 74, 79
　Rainbow, 76
　Taughannock, *68-69*, 74-75, 134-135
　Tinker, 135
　Triphammer, 79
Waterloo, 18
Water parting, *see* Divides
Watkins Glen, 20, 26, 59, *68-69*, 75-77, 80
Wayland, 114
West Danby, 63
West Hill, Ithaca, 94, 127
West River, 21
White Church, 93
Williamsport, Pa., 7, 67, 70, 106
Wilseyville, 64, 91